# Biometric Identification Technologies Based on Modern Data Mining Methods

Stepan Bilan • Mohamed Elhoseny
D. Jude Hemanth
Editors

# Biometric Identification Technologies Based on Modern Data Mining Methods

Springer

*Editors*
Stepan Bilan
State University of Infrastructure
and Technology
Kyiv, Ukraine

Mohamed Elhoseny
College of Computer Information Technology
American University in the Emirates
Emirates, UAE

D. Jude Hemanth
Karunya University
Karunya Nagar, India

ISBN 978-3-030-48377-7 ISBN 978-3-030-48378-4 (eBook)
https://doi.org/10.1007/978-3-030-48378-4



This Springer imprint is published by the registered company Springer Nature Switzerland AG
The registered company address is: Gewerbestrasse 11, 6330 Cham, Switzerland

# Preface

In today's world, the population is constantly increasing. Along with population growth, the number of computing and computer-related tools accompanying people is also increasing. Modern society is characterized by wide automation of various spheres of human activity, and in these processes, a special place is occupied by biometric identification of a person. Biometric identification technologies are everywhere. For example, modern smartphones contain the option of user access through fingerprints, as well as other human biometric parameters. They are also used in the transport sector, automated computer systems, various access systems, forensic science, and other spheres of human life.

As requests for the use of biometric identification tools increase, the number of financial investments in developing new methods and tools with increased reliability of identification and authentication is also increasing. In addition, much attention of specialists is paid to the search for new, previously unknown biometric characteristics, such as accuracy, speed, and ease of implementation. At the same time, the bulk of the methods and means of biometric identification are device-dependent. They require a special interaction of the person's location in relation to scanning and reading devices of a special design. In this regard, researchers' efforts are spent on the creation and development of methods and tools that do not require distraction of a person from his main activity at the time of identification.

This book is very relevant and aims to highlight new modern research in the field of biometric identification. The book will help specialists gain new knowledge and determine the future direction of scientific research.

The book consists of 12 chapters.

The first chapter describes the application of the triangulation method in the process of biometric identification by an image of a person's face. In the second chapter, based on the existing methods, a method for correctly describing a person's gait is presented, and its effectiveness is evaluated on three data sets. The third chapter describes the system of biometric identification by keyboard dynamics. Algorithms for the formation of characteristic features and user identification are considered. The fourth chapter focuses on the description of biometric identification methods by the dynamics of handwritten handwriting. Two approaches to identification are considered. The fifth chapter discusses the method of biometric identification of a person's gait based on parallel shift technology. The sixth chapter describes the use of the Radon transform for biometric identification of a person from handwritten images. The seventh chapter presents a comparison between different detectors (Harris, FAST, SIFT, and SURF) for biometric identification using facial images. The eight chapter discusses the existing problems of face recognition in three-dimensional images. A biometric identification method based on scalar perturbation functions and a set-theoretic subtraction operation is proposed. The ninth chapter describes a new method for analyzing the shape of the palm's inner side. The tenth chapter describes a method for biometric identification of a person by the palm's geometric shape based on parallel shift technology. The eleventh chapter provides an unlimited multimodal biometric system for smartphones, which uses a multiuser selfie and behavioral biometrics. The twelfth chapter describes research in the field of biometric identification based on image analysis of the auricle.

Kyiv, Ukraine — Stepan Bilan
Emirates, UAE — Mohamed Elhoseny
Karunya Nagar, India — D. Jude Hemanth

# Introduction

This book is a research book that can be read by students, analysts, scientists, and developers in the field of data mining and biometric identification and authentication. The book acquaints students and novice practitioners with the basics of biometric personality identification using static and dynamic biometric characteristics, and to highly qualified specialists and researchers in this field the book helps in modern research and new methods. The book is also useful for organizations involved in developing modern devices designed for biometric identification of people in various access systems and other public places. The book is written in a simple and understandable language, describing modern research and new approaches to biometric identification of a person. Based on the modern development of biometric technologies, each library and organization involved in the development of new and reliable biometric identification systems will be interested in acquiring this book.

# Contents

# Possibilities of Applying the Triangulation Method in the Biometric Identification Process

**Muzafer Saračevič, Mohamed Elhoseny, Aybeyan Selimi, and Zoran Lončeravič**

## 1 Introduction

Biometrics is an integral part of every individual. It refers to metrics related to body measurements and calculations which have unique characteristics and differs from each other. It is known that there are no two people whose biometric data could coincide. By applying biometrics, you do not need to remember passwords, PIN codes, etc. Some of the essential and significant characteristics of biometrics are a high level of data security, faster data access, reduction of administrative costs, more efficient keys and confidential data distribution, etc.

The goal of multi-factor authentication is to form a layered defense that relies on two or more different ways of determining the authenticity of the user when logging onto a system (authentication) and giving access privileges (authorization). These techniques combine what the user knows, what the user owns, and what physically identifies the user (biometric verification). The recognition process is based on the authenticity of specific biometric characteristics, whether physical or behavioral characteristics, with those biometric data that are in the database. The most important thing in building a biometric system is the way in which recognition will be based.

M. Saračević (✉)
Department of Computer Sciences, University of Novi Pazar, Novi Pazar, Serbia
e-mail: muzafers@uninp.edu.rs

M. Elhoseny
College of Computer Information Technology, American University in the Emirates, Emirates, UAE
e-mail: mohamed_elhoseny@mans.edu.eg

A. Selimi
International Vision University, Gostivar, North Macedonia
e-mail: aybeyan@vizyon.edu.mk

Z. Lončeravić
ITS - Information Technologies School, Belgrade, Serbia



S. Bilan et al. (eds.), *Biometric Identification Technologies Based on Modern Data Mining Methods*, https://doi.org/10.1007/978-3-030-48378-4_1

In this chapter, the authors consider some possibilities of applying triangulation techniques for obtaining biometric data and converting them into digital form. In addition, also authors consider some processing techniques, which will conduct quality control and create a biometric template, as well as comparison techniques, which compile the obtained biometric templates on the reader with those stored in a particular database.

The chapter consists of seven parts. In Sect. 2, some of the author's previous research in the field of computational geometry, respectively triangulation of the polygons, as well as some research related to the field of cryptography and biometrics are mentioned. In Sect. 3 of this chapter, the biometric authentication system with an emphasis on face recognition technology is described. Section 4 lists triangulation techniques and faces recognition algorithms with an emphasis on face recognition techniques. Section 5 describes the specific system requirements of the proposed solution and demonstrated the basic phases of the proposed system for the biometric verification process. This process consists of the following phases: taking a biometric printout, template creation, comparing results, and decision making phases. Section 6 of this chapter lists the successful tests in the authentication process. In particular, a proposed solution based on the application of the triangulation method in combination with the face recognition technique using the appropriate two sets of data. At the end of the chapter authors list concluding observations and suggestions for further research in the field of biometric authentication systems.

## 2 Previous Research

The author's previous paper [1], presents a procedure for the application of computational geometry in the process of generating hidden cryptographic keys from 3D image. In paper [2], the authors describe some authentication techniques with an emphasis on face recognition technologies. The author's paper [3] describes one method of recording triangulations of a polygon. The obtained notation authors express in the form of binary records. A presented method of storing triangulations in this study is based on Lukasiewicz's algorithm and binary trees. Particularly, authors provide Java implementation for binary notation of triangulations.

Authors in [4] examine a unique aspect of the development of new approaches and methodologies for biometric identification and verification. The use of recently developed advanced techniques in computational geometry and image processing is examined with the purpose of finding the common denominator between the different biometric problems and identifying the most promising methodologies. In [5] authors show the advantages of recognition of a fingerprint based on triangle parameters from the Delaunay triangulation of minutiae.

Also, authors in [6, 7] present a novel method for fingerprint minutiae matching, where the proposed method using a Delaunay triangulation. This method based on the relative position and orientation of each minutia with respect to its neighbors obtained by the triangulation. The minutiae triangles have good discrimination power since, among all possible minutiae triangles, they are the only ones satisfying the properties of the Delaunay triangulation. This is an important feature.

## 3 Biometric Authentication System Based on Face Recognition Techniques

Face recognition is a very important area which is used in the identification of objects with special application. Identification and authentication methods have become a key technology in various segments of modern society, such as controlling entry to buildings and premises, controlling access to computers and other computer systems, as well as in the field of a criminal investigation. With modern face recognition, the degree of identification is achieved even over 98%, under well-controlled conditions of lighting, motion, and pose [8–11]. In the modern IT era, where computers become the basic tool in daily work processes, the new authentication tasks appear increasingly. Together with the ability of "understanding" of movement and face lifting, recognition as technique will make possible intelligent interaction between man and machine. Face recognition from a series of images presents a special challenge for researchers, both because of the greater reliability of the method and the complexity of recognition.

In the last decade, face recognition using computers has intensified considerably, and recently a special emphasis has been placed on recognition from the mobile image, i.e. the ever-increasing use of video surveillance in all segments of social life. The biggest problem in recognition is in the image looking which depends on the angle of view, lighting, etc. Namely, 3D objects, such as a human face, can take many different expressions when projected onto a 2D image. Another problem is the content of the scene or background elements that can be overwhelmed by different objects and it is very difficult to extract the area of the image containing the face. It is possible to cover the objects so only a part of the information will be available to the recognition system. In [12], the authors show 3D adaptive tomography using Delaunay triangulation and Voronoi polygons. The authors give an algorithm for an automatic regrinding, which is based on Delaunay-Voronoi tessellation. It increases the local pixel density where the null space energy is low or the velocity gradient is large and reduces it elsewhere. In paper [13] authors present the application of Delaunay triangulation to face recognition. In [14] authors show a faceprint technique based on the local feature. A new approach that differentiates from the conventional face recognition techniques in partitioning the facial image into a triangular feature area is proposed. A Delaunay triangulation based on barycenters replaces triangles. The result of employing this Delaunay triangulation method is invariant to scale and orientation differences of the facial image tested, and robust to the contamination of noises.

The facial recognition problem can be formulated in a way that from the image or video identifies one or more persons which are in the available facial database. The database can also contain some other data (such as age, sex, skin color) that can help in selection in the searching. Solving the problem of recognition is not at all a simple business and involves several phases:

1. separating faces from a complex scene,
2. studying facial recognition features (feature extraction)
3. comparison with the available data in the database for the purpose of identification
4. methods—include additional processing steps such as, for example, estimation of the angle of observation, assessment of facial expression, etc.

Facial recognition from a static or dynamic image with a computer becomes an active research area that extends to several disciplines such as image processing, pattern recognition, computing vision, and neural networks. The technology for face recognition has a large number of applications in various security and commercial systems: from statically comparing two controlled-aspect photos (such as passport photos) to real-time video recognition from video sequences (e.g., video surveillance cameras).

Recognition based on a photo from personal documents or an automatic search from criminal evidence is a typical example of a comparison based on a static image. These photos are characterized by an acceptable light control level, good image quality and specific background type, camera resolution and the distance between the camera and the person which is photographed. The existence of these standards during image capture makes it easier to extract face from the image and application of matching algorithms. A potential challenge in this issue is the search for a large set of images as well as just matching.

Security surveillance systems (using single images) are more complicated in terms of face recognition due to an uncontrolled background appearance, which may include a range of other objects. Therefore, separating the face from the image becomes more difficult. Each person has a set of unique physical characteristics, which can be measured and compared with each other, using appropriate technologies. In paper [5] authors show fingerprint recognition for securing the transaction. Also, the authors propose features based on triangle parameters from the Delaunay triangulation of minutiae.

## 4 Triangulation Method and Face Recognition Technique

In the face recognition of different technologies in the faceprint creation, i.e. sampling, are applied, and they are done the comparison of the received and stored data of the person in the database. Key details are measured and numerical code (ie a binary record) is generated, representing a person in the database.

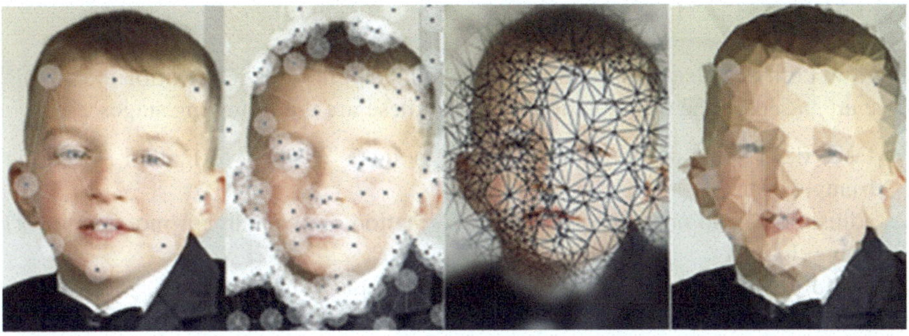

**Fig. 1** The phase of forming a faceprint or template based on the facial triangulation method

As with other biometric identification systems, also the processes of enrollment and verification and identification processes differ from each other. The typing phase usually takes longer than the testing phase, during which several images of the same face are taken. Ideally, images are taken with slightly different facial angles to the camera. After taking pictures, the characteristic features of the face are distinguished and creates a faceprint or template.

Figure 1 shows the procedure for forming faceprint through a certain triangulation, starting from the formation of initial triangulation, then to complex networks, and at the end of generating 3D views that are stored in a binary form in a database.

The procedure of recording polygon triangulations will be connected with the binary trees [3]. It is important to note that the binary tree can represent Catalan numbers. More specifically, based on these numbers, is performed the process of calculating the number of all triangulations of convex polygon [15]. In previous papers [16–18], authors dealt with the polygon triangulation problem and Catalan numbers.

For each triangulation, authors applied the *Lukasiewicz's* algorithm (see papers [3, 15]). In this way authors get the corresponding binary record (notation), which is unique and corresponds to exactly one triangulation for which this binary tree is applicable. In author's papers [1, 3] have explained in detail how on the basis of a binary tree (more precisely, its binary record), a convex polygon triangulation is formed.

From the example shown in Fig. 1, for a part of the chosen triangulation in the image obtain the following 120-bit bipartite faceprint:

```
Faceprint_bin=111101001110010111010100001101110001011101000
111100000100000101111010011100101110101000011011100 0
101110100011110000100001 0
```

In the faceprint database in the form of key is stored the decimal values for each image. In this case, it is obtained the following key:

`Key_decimal= 127158191870200575835247609375755 2706.`

Now, define the basic modules in the technique of the face recognition that in this case are based on the triangulation method. These are the following stages:

1. face detection—definition of initial points for the formation of basic triangulation
2. adjustments and normalizations—the formation of an enlarged (more complex) triangulation
3. encoding only faces—forming a binary file of faceprint
4. comparison and identification phase.

In the detection phase, face recognition software searches for the face in the images that are entered in the test phase, that is, when it comes to identification, then it searches for a shape (triangulation) that matches the face definition. When a shape that resembles a human head is detected, the software switches to the processing stage. In these papers [9, 10], a method is proposed that improves the recognition rate by selecting and generating an optimum image of the face from the empirical facial images.

After the detection phase, the adjustment phase (normalization) is followed, where the position size and orientation of the head are determined. After the normalization phase, it follows the coding phase, which is also an important phase in the face recognition process. In the coding phase, the measurement, i.e. translation of key details from the normalized 2D frontal image of the head into a single digital code, is performed (in this case, *Lukasiewicz's* method of triangulation traversing that applied in author's previous works [1, 3]) is used. The obtained digital code (faceprint) is used in the comparison phase to compare with pre-generated codes in the training phase of the system.

In the papers [19, 20], methods are presented which can with certain precision monitor the facial changes and, based on their realization, assume the appearance of a person at a certain moment or phase of age with fairly reliable generic models. These methods in the creation and projection of age for performance, require the use of a large database with attributes displayed in several stages. Based on this, a model for methods based on age estimation using a classification or by applying regression to data from a database is created.

It is important to emphasize that, as in the case of fingerprint recognition technology and face recognition technology, it is quite vulnerable to the possibility of counterfeiting. In the paper [21], there are ways to deceive in face recognition techniques, how to protect themselves, and how in real-time, without additional devices, it can be determined whether it is an attempt to deceive the system. Some researchers also used authentication of the voice-face, as well as the exploration of the lips, or speech movements. With thermal infrared cameras, it is possible to conclude successfully whether the object is real or it is an attempt to deceive the system [22]. An eye blink allows the detection of false object placement without additional hardware modules.

In an experiment carried out and published by the author in [23], it is concluded that a significant role in the detection and detection of intrusion is the role of two or more dependent frames of the image. Anti-deception is also a challenge for explorers. One problem with this method is the problem if the eyes are not noticeable enough and can not be defined whether they are open or closed.

## 5 Specification of the Proposed Model

Face recognition is actively used in the field of *Computer Vision* as well as in the field of biometrics. The essence lies in the development of applications for security, robotics, human-computer interaction, artificial intelligence, digital cameras, games, and entertainment. Author's GUI face recognition application is developed in Java NetBeans surrounds and includes two modules:

1. *Face detection module*—represents the phase where the face is found in the photo and is processed the image for easier identification of face.
2. *The face recognition module*—represents the phase when a person is found. In this module, the person is detected and processed in relation to the database of recorded persons. Practically it represents the role of the system to decide which person is in the image whose image is being processed or to check the templates (based on recorded triangles and digital faceprints) and the face value on the image (generated triangulation) which is needed in recognition in the relation of data from the database.

### *5.1 Strategy for Categorization of Face Recognition*

Face detection from the 2D image with the appropriate algorithm for detection and face recognition makes the technique implemented in the proposed solution. In the paper [24], is considered how a digital face map is created from the 2D image at the input, and then follows the recognition based on the stored data in the database.

It is important to note that there are different schemes and strategies for categorizing face recognition. Holistic methods try to recognize the face in terms of the whole image, or the face as a whole. On the other hand, non-holistic approaches are based on the identification of certain facial features, such as the nose, mouth, eyes and their relationship, in order to make the final decision based on it. Some methods try to take advantage of both approaches at the same time and therefore they are classified as hybrid. In this case, the strategy includes the following stages:

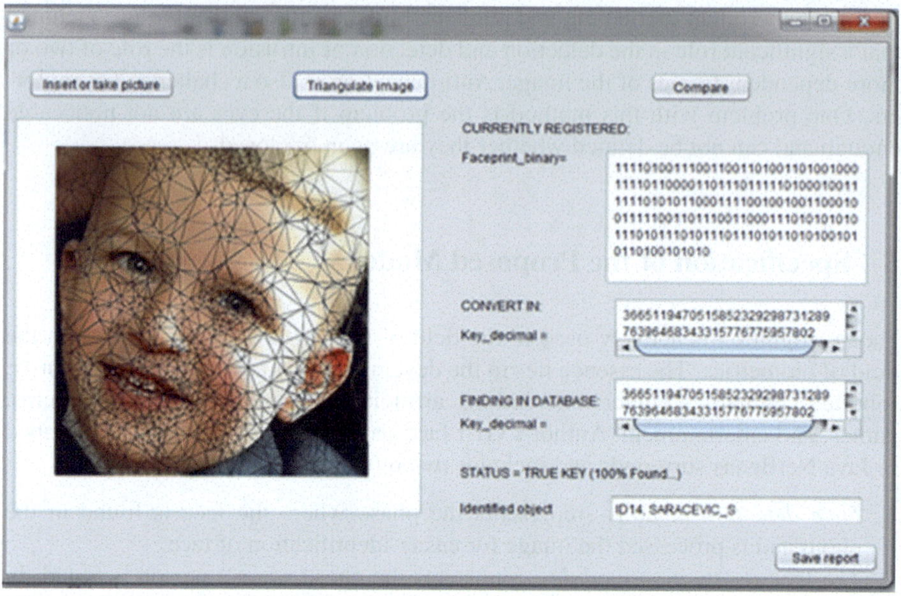


**Fig. 2** Comparison and identification phase based on faceprint from the database

1. *Face detection*—the geometric characteristics of the face, individual parts of the face, such as eyes, nose, and lips, are taken.
2. *Forming a digital map*—the corresponding algorithm provides a set of values characteristic for the given face.
3. *Recording a digital map in a folder and keeping it in the database*—in order to be able to recognize a person, it is necessary that the face is stored in the database, in a defined format.
4. *Face recognition*—is done by comparing the individual with the one stored in the database, all based on the corresponding recognition algorithms.

Figure 2 shows one possible technique for identifying faces from the image. Faceprint in the form of triangulation is loaded from the database in the form of key_decimal, after which this value was generated in Faceprint_binary. After that, authors are loading another image, triangulating and generating another Faceprint_binary key.

If it matches the comparing procedure with the loaded record from the database then the authentication has the status TRUE.

## 5.2 User Requirements

Basic user requirements are:

- To the user is enabled record, each record about the user represents the basic information about it, that is, allows the user name entry.
- The system accepts the image at the input, creates a template and enables it to save it as comparing and displaying the results.

- The system allows you to load data from the face database,
- The system allows testing in the system more:more.
- The system enables monitoring the recognition process and recording the results.

## *5.3 System Requirements*

Basic system requirements are:

- The user expects the system to display the recognition results for it.
- Cases of use:
  - Image input—The user can enter the image and additional data into the database.
  - Comparison—The user requires a comparison with the stored data.
  - Display results—Displays the results of the comparison (with "save report").
- Event stream:
  - Basic stream: The user input the name, the user allows the system to import pictures, the user starts testing.
  - The system performs testing: a comparison and displays the result.
  - Objective: The user has been identified as a "Person_name".

## *5.4 General Scenario for 2D Face Recognition*

It is important to note that no detection algorithms should be mixed with the one for recognition. The overall efficiency of this authentication system depends on both of them. It should be noted that there are 2D and 3D algorithms that are implemented together and which together make face recognition systems.

Figure 3 shows the general scenario of the proposed solution based on 2D face recognition.

## *5.5 Advantages of the Proposed Model*

The key advantages of this model are:

- Each person has his own key on the basis of the received digital code or faceprint (since it is derived from triangulation, it can be a Catalan number or Catalan-key).
- This way provides secure transmission (distribution) of keys.
- The space of such keys is extremely large (see [1, 17] in which authors analyzed the complexity of the space of Catalan keys).

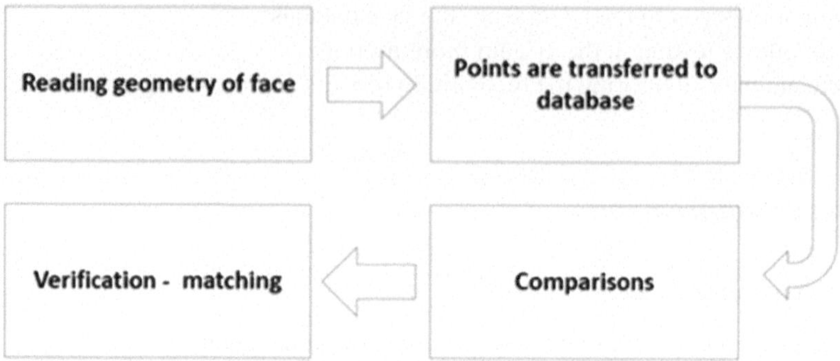


**Fig. 3** Simple 2D face recognition scheme

In paper [25], authors describe an algorithm for generating 3D models of human faces from uncalibrated images. Their method based on locally applying Delaunay triangulation method. In the paper [26], the authors presented a relatively new model for 3D face recognition, i.e. reconstruction and obtaining 3D shapes from only one image. This method uses global facial similarities by combining shading and shape information with generic information derived from a single reference model. This method allows the creation of a clear 3D image from traditional 2D images, such an algorithm has better performance [11].

The extension in the form of the proposed method for 3D face recognition based on the triangulation process consists of four modules: *Face detection, feature search, final feature detection, triangulation* (see Fig. 4). Phases like *Warping* and *Feature extraction* represent the verification process or the matching module.

The "Feature search" module distinguishes features such as nose, whole, mouth, and receives an optimized 3D surface for further extraction of biometric patterns [27–29]. During the biometric template extraction, two phases are used:

1. In the first phase, the surface "semantic analysis" was performed, resulting in the location of the key (crania-facial) features to generate generic topological maps of the face.
2. In the second stage, when the location of the specific surface of the face is known (eyes, specific arches, forehead, nose, mouth, beard zone, etc.), information about the local surface characteristics is distinguished.

In Fig. 5, the result of the "Feature search" module is displayed in the process of separating the mark in the first case from the frontal image, and the other two cases are related to the images where the face is painted from different angles.

Appropriate methods that work in the identification mode compute separate biometric templates with all the stored templates in the database and produce a similarity with estimation from the stored templates. This information is further used to

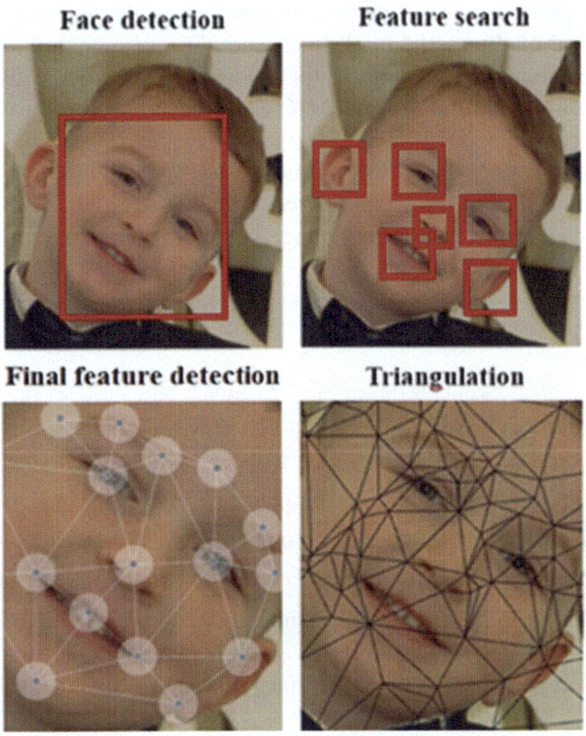


**Fig. 4** Phases in 3D face recognition based on the triangulation method

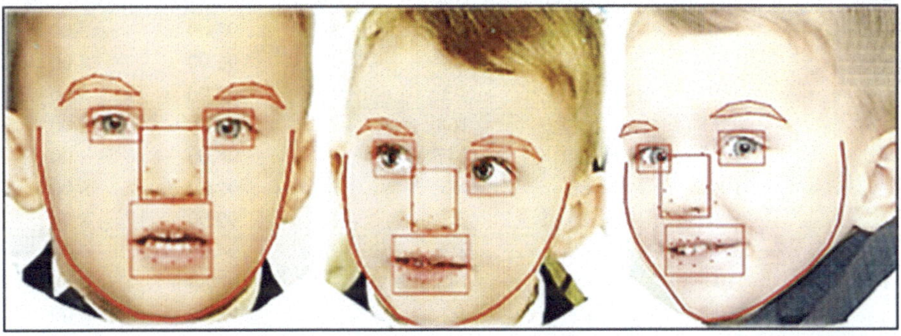

**Fig. 5** "Feature search" option from the frontal image and two images from different angles

build a geometric descriptor and pack these data into a biometric print. The output module is a biometric template that features a unique face, which is used in the matching phase.

The main technological limitation of the 3D face recognition method is the acquisition of a 3D image, which usually requires a series of cameras. This is also the reason why 3D facial recognition and their methods significantly later appeared in relation to 2D methods. One of the methods of automatic 3D face recognition published in papers [30, 31] based on the method for determining the range of images with the *Surface Interpenetration Measure* (*SIM*) as a measure of the similarity between two facial images.

The results of authentication were obtained by combining the *SIM* values corresponding to the matching of four different regions of faces such as the circular and elliptical region around the nose, forehead and whole face.

## 6 Experimental Results

In this section of the chapter, authors tested a proposed solution based on the application of the triangulation method using the appropriate two sets of data:

1. The first set is used in the system training (in the Table 1 called "training set")
2. The second set is a photographic database of the person to test the system (in the table called "test set").

The training set consists of randomly selected facial images that make up 70% of the faces of the database. For the needs of the training session are generated the files that contain the path to the order of 150, 100, 50 and 25 face images, which are also

**Table 1** Testing results from the database system for the triangulation method

| TEST | Training set | Test set | True | False | True (%) | False (%) |
|---|---|---|---|---|---|---|
| T1 | 150 | 100 | 97 | 3 | 97.00 | 3.00 |
| | 150 | 50 | 48 | 2 | 96.00 | 4.00 |
| | 150 | 25 | 23 | 2 | 92.00 | 8.00 |
| | SUM | 175 | 168 | 7 | 95.00 | 5.00 |
| T2 | 100 | 150 | 147 | 3 | 98.00 | 2.00 |
| | 100 | 50 | 46 | 4 | 92.00 | 8.00 |
| | 100 | 25 | 23 | 2 | 92.00 | 8.00 |
| | SUM | 225 | 216 | 9 | 94.00 | 6.00 |
| T3 | 50 | 150 | 143 | 7 | 95.33 | 4.67 |
| | 50 | 100 | 92 | 8 | 92.00 | 8.00 |
| | 50 | 25 | 22 | 3 | 88.00 | 12.00 |
| | SUM | 275 | 257 | 18 | 91.78 | 8.22 |
| T4 | 25 | 150 | 139 | 11 | 92.67 | 7.33 |
| | 25 | 100 | 85 | 15 | 85.00 | 15.00 |
| | 25 | 50 | 41 | 9 | 82.00 | 18.00 |
| | SUM | 300 | 265 | 35 | 86.56 | 13.44 |
| T1–T4 | SUM (1–4) | 975 | 906 | 69 | 91.83 | 8.17 |

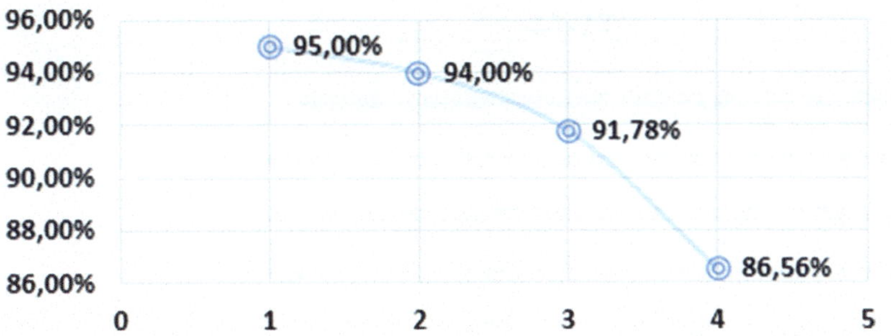


**Fig. 6** Trend of positive recognition rate

selected by randomly from the sample by using the automated tool generated for this test.

Based on the results obtained from tests 1 to 4, it can be noted that with a decrease in the number of people in the training phase of the system with an initial 95% performance (a training set of 150 images), the recognition performance is reduced to 86.56% of the accuracy of recognition during system training (training set of 25 images). Therefore, for a more successful degree of recognition, it is necessary to train a system with as many images as possible, but there is also one problem here. Namely, in too many training images, the performance of the system in terms of computer time for processing is rapidly decreasing, so sometimes the waiting time of the system sometimes spends even a few minutes.

In Fig. 6, presents the tendency of falling recognition rates when the face images are reduced in the training set.

Finally, from the analyzed individual test results, the results are combined and systematized to show the overall efficiency of the system based on the application of the triangulation method.

Figure 7 shows the performance of the system for four tests individually (T1–T4), and the final average performance of the system for the mentioned technique is approximately 92%.

Thus, generally speaking, using the triangulation method in combination with face recognition technique, the authentication performance rate is equally successful as recognition with the application of "*Mahalanobis distance*" or "*Euclidean distance*" [24].

In some further research and improvement of proposed method, authors can also discuss combined techniques such as "Euclidean Minimum Spanning Tree Problem" [32, 33] or "Euclidean Traveling Salesperson Problem" on the Delaunay triangulation. In paper [34], authors present a fast approach to solving large-scale traveling salesman problems (TSP) based on Delaunay Triangulation. In paper [35], authors give several motivations for Delaunay and greedy triangulations. Indeed, their experiments show that the resulting tours are in average within 0.1% of the optimality.

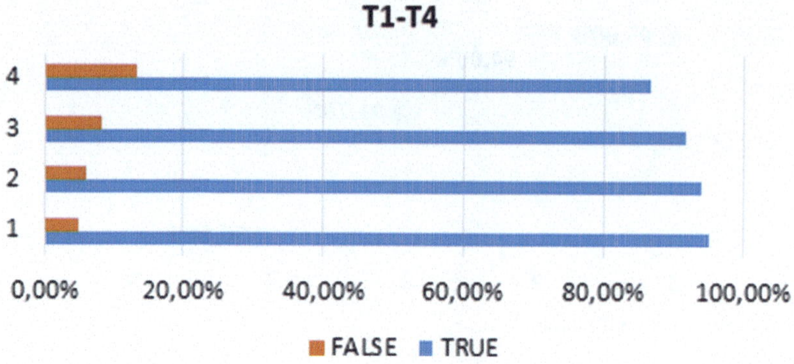


**Fig. 7** The share of positive and negative recognition rate

## 7 Conclusion and Further Works

The biometric identification process based on face recognition in the future will get importance. The emergence of new technologies (such as a high-resolution camera and then the emergence of 3D cameras, and increasingly sophisticated computer technologies) will only make this technique significant in both the identification process and other areas of human life. Biometric identification process in combination with other types of biometrics, in the form of multimodal systems, it creates an authentication technique that adjusts of various aspects and hybrid systems (integration with voice recognition, fingerprints, or palm geometry, as well as recognition of iris and retina of the eye).

In this chapter, the authors presented the possibilities of applying the triangulation method in the biometric identification process. A proposed method for authentication is based on face recognition technologies and polygon triangulation as a fundamental algorithm in computational geometry and graphics. The results of the experimental part show that using the triangulation method in combination with the face recognition technique, the success rate of authentication is achieved equally well as recognition with the application of some other methods.

Further scientific research and the future of biometric systems relate to the integration with artificial intelligence as well as performance improvements. Also, authentication techniques based on video sequencing algorithms will have significant advantages in law enforcement and traffic control. The future of the face recognition based authentication technique lies in the methods that use face recognition with a focus on 3D technology. Increasing using artificial intelligence can also contribute to the efficiency of this method, as well as improving performance. A large number of published papers and the research is explained about the enviable progress of face recognition through the evaluation of algorithms and techniques for the very implementation of such systems. Also, in many published papers, successful facial recognition exceeds 95% of the accuracy, depending on the algorithm, technology, and the very way of research.

It is possible to observe that with simple implementation and technology, with minimal costs compared to other biometric authentication systems, the recognition face is not lagging behind, and in many cases, it exceeds other techniques in the achieved results. According to research by Lawrence Hornack, director of the Center for Identification of Identity Technologies (CITeR), in the future, cameras will read the structure of the eye and shape of the person without waiting for a special spot for checking or watching in camera. More specifically, passengers will be identified in the process of walking in corridors of buildings, airports, etc. So, the future of authentication lies in facial recognition methods.

## References


1. Saracevic, M., Selimi, A., & Selimovic, F. (2018). Generation of cryptographic keys with algorithm of polygon triangulation and Catalan numbers. *Computer Science, 19*(3), 243–256.
2. Loncarevic, Z., Saracevic, M., & Hadzic, M. (2018). Analysis of some techniques of biometric authentication systems with emphasis on 2D and 3D face recognition technology. *Economic Challenges: Ekonomski izazovi, 7*(13), 119–139.
3. Saracevic, M., Masovic, S., & Milosevic, D. (2013). Java implementation for triangulation of convex polygon based on Lukasiewicz's algorithm and binary trees. *Southeast European Journal of Soft Computing, 2*(2), 40–45.
4. Gavrilova, M. (2007). Computational geometry and image processing in biometrics: On the path to convergence. In *Image pattern recognition: Synthesis and analysis in biometrics* (pp. 103–133). Singapore: Word Scientific.
5. Vibert, B., Le Bars, J. M., Charrier, C., & Rosenberger, C. (2017). *Fingerprint class recognition for securing EMV transaction.* In International conference on information systems security and privacy.
6. Parziale, G., & Niel, A. (2004). A fingerprint matching using minutiae triangulation. In *Biometric authentication - lecture notes in computer science* (Vol. 3072). Berlin: Springer.
7. Bebis, G., Deaconu, T., & Georgiopoulos, M. (1999). Fingerprint identification using delaunay triangulation. In *Proceedings on international conference on information intelligence and systems.*
8. Delac, K., Grgic, M., & Stewart, B. M. (2008). *Recent advances in face recognition.* Croatia: In-Tech.
9. Sirovich, L., & Meytlis, M. (2009). Symmetry, probability and recognition in face space. *Proceedings of the National Academy of Sciences, 106*(17), 6895–6899.
10. Kim, D., & Sung, J. (2009). *Automated face analysis: Emerging technologies and research.* New York: Digital Media Research Lab, LG Electronics.
11. Ding, L., & Martinez, A. M. (2010). *Features versus context: An approach for precise and detailed detection and delineation of faces and facial features.* Columbus, OH: Ohio State University.
12. Böhm, G., Galuppo, P., & Vesnaver, A. (2000). 3D adaptive tomography using Delaunay triangles and Voronoi polygons. *Geophysical Prospecting, 48*, 723–744.

13. Chiang, J. Y., Wang, R. C., & Chang, J. L. (1997). The application of delaunay triangulation to face recognition. In *Proc. national computer symp, TaiChuang, Taiwan* (Vol. 1, pp. 27–32).
14. Zhang, G., & Wang, Y. (2009). Faceprint: Fusion of local features for 3D face recognition. In *Advances in biometrics - Lecture notes in computer science* (Vol. 5558). New York: Springer.
15. Koshy, T. (2009). *Catalan numbers with applications*. New York: Oxford University Press.
16. Saracevic, M., & Selimi, A. (2019). Convex polygon triangulation based on ballot problem and planted trivalent binary tree. *Turkish Journal of Electrical Engineering and Computer Sciences, 27*, 346–361.
17. Stanimirović, P., Krtolica, P., Saracevic, M., & Masovic, S. (2014). Decomposition of catalan numbers and convex polygon triangulations. *International Journal of Computer Mathematics, 91*(6), 1315–1328.
18. Selimi, A., & Saracevic, M. (2018). Computational geometry applications. *Southeast European Journal of Soft Computing, 7*(2), 8–15.
19. Zhang, Y., & Zhou, Z. H. (2010). *Cost-sensitive face recognition*. Nanjing, China: Nat. Key Lab. for Novel Software Technol. Nanjing Univ..
20. Kumar, R., Barmpoutis, A., Banerjee, A., & Vemuri, B. C. (2011). *Non-Lambertian reflectance modeling and shape recovery for faces using tensor splines*. Gainesville, USA: Dept. of Comput. & Inf. Sci. & Eng., Univ. of Florida.
21. Suo, I., Zhu, S. C., Shan, S., & Chen, X. (2010). *A compositional and dynamic model for face aging*. Beijing, China: Grad. Univ. of Chinese Acad. of Sci..
22. Xue, Z., Ming, D., Song, W., Wan, B., & Jin, S. (2010). *Infrared gait recognition based on wavelet transform and support vector machine*. Tianjin, China: College of Precision Instruments and Optoelectronics Engineering, Tianjin University.
23. Phillips, P. J., Scruggs, W.T., O'Toole, A. J., Flynn, P. J. (2007). FRVT 2006 and ICE 2006 large-scale results. NISTIR 7408 - National Institute of Standards and Technology, USA.
24. Lončarević, Z. (2011). Authentication techniques based on face recognition. In *Master thesis: Faculty of Organizational Sciences*. Belgrade, Serbia: University of Belgrade.
25. Hassanpour, R., Atalay, V. (2004). Delaunay triangulation based 3D human face modeling from uncalibrated images. In *Conference on computer vision and pattern recognition workshop*, Washington, USA (p. 75).
26. Queirolo, C. C., Silva, L., Bellon, O. R. P., & Segundo, M. P. (2010). *3D face recognition using simulated annealing and the surface interpenetration measure*. Curitiba, Brazil: Univ. Fed. do Parana.
27. Park, U., Tong, Y., & Jain, A. K. (2010). *Age-invariant face recognition*. East Lansing, MI: Dept. of Comput. Sci. & Eng., Michigan State Univ..
28. Berretti, S., Del Bimbo, A., & Pala, P. (2010). *3D face recognition using iso-geodesic stripes*. Firenze, Italy: Dipt. di Sist. e Inf. Universitd degli Studi di Firenze.
29. Wang, Y., Liu, J., & Tang, X. (2010). *Robust 3D face recognition by local shape difference boosting*. Hong Kong, China: Dept. of Inf. Eng. Chinese Univ. of Hong Kong.
30. Kemelmacher-Shlizerman, I., & Basri, R. (2011). *3D face reconstruction from a single image using a single reference face shape*. Seattle, USA: Dept. of Comput. Sci. & Eng., Univ. of Washington.
31. Liu, I., Chen, Y., & Tang, X. (2011). *Decomposition of complex line drawings with hidden lines for 3D planar-faced manifold object reconstruction*. Hong Kong, China: Dept. of Inf. Eng. Chinese Univ. of Hong Kong.
32. Keil, J. M., & Gutwin, C. A. (1989). The Delaunay triangulation closely approximates the complete Euclidean graph. In *Algorithms and data structures - Lecture notes in computer science* (Vol. 382). New York: Springer.

33. Sanguthevar, R. (2005). On the Euclidean minimum spanning tree problem. *Computing Letters, 1*(1), 11–14.
34. Evgeny, Y., Eckart, S., Ulrich, S., & Kurt, T. (2005). An O(n log n) Heuristic for the Euclidean traveling salesman problem. Research Report Series on University of Würzburg. Institute of Computer Science, Germany.
35. Letchford, A. N., & Pearson, N. A. (2008). Good triangulations yield good tours. *Computers & Operations Research, 35*(2), 638–647.

# Biometric Gait Identification Systems: From Spatio-Temporal Filtering to Local Patch-Based Techniques

**Mohammad H. Ghaeminia, Shahriar B. Shokouhi, and Abdollah Amirkhani**

## 1 Introduction

Human's gait or style of walking is considered to be an effective biometric measure for identifying the individuals in public places for surveillance programs [1]. This is mainly attributed to the fact that the images of the walking can be easily acquired from a distance. Nevertheless, any biometric system based on human's gait identification can be affected by certain external factors such as the type of clothing, bagging condition, camera viewpoint, surface, and the aging [1, 2]. Despite these challenges, it is important to note that the manner of walking can still be relied upon as a unique biometric benchmark for solving identification problems [3]. The key issue in handling a gait under covariate factors is developing an efficient biometric system. In this paper, we propose the criteria for ranking the gait biometric systems and evaluating them accordingly.

Basically, the most common way to describe the gait without a predefined model is using template feature [1, 3–5]. For example, gait energy image (GEI) [1], general tensor discriminant analysis (GTDA) [6], gait flow image (GFI) [7], chrono-gait image (CGI) [8], and 5/3 gait image (5/3GI) [3] are some of the template features that have been developed by the researchers. In such methods, the sequence of silhouettes (or features) is aggregated into a single image called "template." Since the nature of the gait is a spatio-temporal process, such conversion has three main limitations [9, 10]: (1) removing temporal ordering of gait, (2) utilizing not an efficient human's motion model, and (3) aggregating gait defects in the template feature.

M. H. Ghaeminia · S. B. Shokouhi
Department of Electrical Engineering, Iran University of Science and Technology, Tehran, Iran
e-mail: mhghaeminia@elec.iust.ac.ir; bshokouhi@iust.ac.ir

A. Amirkhani (✉)
Department of Electrical Engineering, Iran University of Science and Technology, Tehran, Iran
e-mail: amirkhani@iust.ac.ir; amirkhani@ieee.org

 
S. Bilan et al. (eds.), *Biometric Identification Technologies Based on Modern Data Mining Methods*, https://doi.org/10.1007/978-3-030-48378-4_2

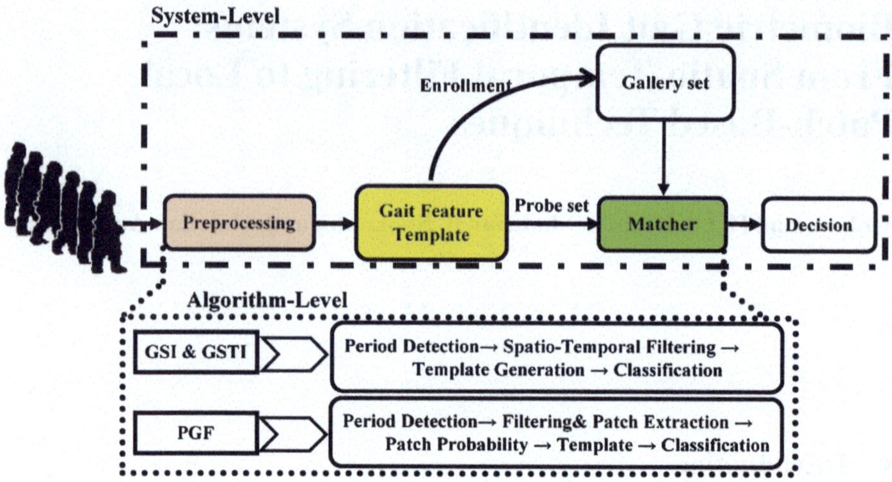


**Fig. 1** The main components of a gait biometric system

In order to resolve these issues, three separate algorithms have been proposed in this paper: (1) gait spatial image (GSI) features, (2) gait spatio-temporal image (GSTI) features, and (3) patch gait features (PGF). Each of these algorithms has a certain advantage in describing the rhythm of human walking; which will be examined in this paper. Besides the efficacy of the above methods in describing the gait, applying them in a real gait biometric system will be limited. Figure 1 shows different between the algorithm-level and system-level evaluations.

As shown in Fig. 1, a method of gait recognition is a subset of biometric system and hence, other benchmarks rather than recognition accuracy should be considered for evaluating the system. In this paper, the performance of gait biometric system is measured according to three proposed benchmarks: (1) recognition error rate, (2) computational and memory complexity, and (3) scalability.

Considering the recent state-of-the-art methods in gait recognition [3, 5, 9–11], the main contribution of the paper is as follows:

- Algorithm-level evaluation of the recent algorithms on three well-known datasets (USF [12], CASIA-B [13] and OU-ISIR-B [14]).
- Discussing on the benchmarks for validation of a gait biometric system.
- Evaluation of the proposed biometric systems according to given benchmarks.
- Providing an efficient tool for measuring the quality of gait system.

The rest of the paper is organized as follows: In next section we discuss on related gait recognition methods. The proposed algorithms (i.e., GSI, GSTI, and PGF) are explained shortly in Sect. 3. The proposed benchmarks for validating a gait biometric system are described in Sect. 4. Meanwhile, Sect. 5 provides the experimental results and discussion and the paper is concluded in Sect. 6.

## 2 Similar Works

The basis of efficient gait identification algorithms is gait template identification by means of template features [3, 5, 7, 15–18]. Therefore, the effective performance of a biometric system strongly depends on the type of algorithm used and the features selected. However, using appropriate features that are robust against the challenging covariates of human gait is itself a basic challenge. Moreover, some algorithms, despite being highly accurate in gait identification (e.g., [4, 19, 20]), have a large computational load and thus limited chance of development in a biometric system.

For appropriate gait identification, it is necessary to compute spatio-temporal features in one period and to compile them in a single image. This subject was investigated by Wang et al. [8] who presented a color template called chrono-gait image (CGI). Also, by extracting 5/3 wavelet features from silhouettes, Atta et al. [3] have recently proposed the 5/3 gait image (5/3GI) template, which not only has a small computational load but also a high gait identification accuracy. Fendri et al. [21] have dynamically selected the different sections of a silhouette which conform to gait covariates and which use the semantic classification scheme to identify different parts of human body. For detecting the gait template that matches human gait template, an effective spatio-temporal filter has been developed [9, 10, 22]. Recent research works have confirmed the existence of certain prominent regions in human gait (i.e., patches) that contain important information [11, 17]. By identifying these regions and describing human gait template on the basis of patches, gait identification accuracy can be increased.

From system perspective, as was stated earlier, there are certain limitations in using the above approaches in a biometric system. In other words, in system-level comparison, the measure of superiority is not just the accuracy of a particular method. For quantitative comparison between various systems, different methods have been developed; mostly on the basis of the performance of face and fingerprint biometrics [23–25]. For example, Grother et al. [23] and Olsen et al. [26] have developed standard ranking criteria for fingerprint biometric systems based on the quality of input sample. Also, Fernandez et al. [25] have examined and developed different evaluation measures, such as system error, in addition to the input sample quality. However, in gait identification, because of using noisy images recorded from a far distance, it is not a good idea to perform evaluations based on the quality of input sample. An appropriate criterion for evaluating the performances of different systems is to evaluate them based on system error [23, 25]. In addition to system error, two more criteria (i.e., computational load and scalability) will also be used in this paper to compare the performances of gait identification systems. The three selected approaches of GSI, GSTI, and PGF will be examined briefly in the next section and then the selection criteria will be thoroughly explained in Sect. 4.

# 3 Selected Approaches

Since human gait is essentially a spatio-temporal process, its proper description should be based on spatio-temporal features. For this purpose, a template feature based on spatio-temporal features has been developed in recent years. Of these methods, the three approaches of GSI [10], GSTI [9], and PGF [11] have been able to properly exploit the properties of corresponding features. In this section, the three cited methods will be introduced briefly and then the most effective classification technique for gait identification will be explained. It should be mentioned that the materials stated in this section pertain to algorithm-level development.

## *3.1 Preprocessing of Data*

For the proper extraction of features, preliminary processing of information should be carried out on raw video recordings. The main preprocessing steps are as follows: resolving the background, extracting the foreground, and computing the gait period [10]. The Gaussian mixture model (GMM) has been proposed for modeling the background and subtracting the foreground from it [12]. Foreground images are obtained by subtracting the gait sequence from the background model. Following the extraction of foreground images, these images are normalized and made uniform by considering the obtained image centers [12]. In this way, silhouette images (foreground) with identical sizes and aspect ratios will be obtained. Then the gait period can be calculated by counting the pixels in the lower half of the silhouette image in each frame [10]. Because the values of corresponding pixels, when a walker's two feet move away from or get closer to each other, will have local maxima and minima. By computing the local extremums, the gait period (*T*) is easily obtained by considering the average distance between two successive maxima or minima.

## *3.2 Gait Spatio-Temporal Filtering*

As we know, the sequence of human gait is a type of spatio-temporal signal in 3-D space, where $x$ and $y$ are two spatial coordinates and $t$ is the temporal coordinate. Since there is no displacement in the $y$ direction, in the GSI approach, the gait sequence has been considered in the $x$-$t$ space. For this purpose, a directional filter has been developed for determining the direction of gait in this space [10]:

$$\begin{cases} Dir,L_1 = R_{A1} - R_{B2}; & Dir,L_2 = R_{A2} + R_{B1} \\ Dir,R_1 = R_{A1} + R_{B2}; & Dir,R_2 = R_{A2} - R_{B1} \end{cases}. \quad (1)$$

In the above equations, $Dir_{,}L_i$ and $Dir_{,}R_i$ ($i = 1, 2$), respectively, denote the main kernels for determining gait directions to the left or right in two different strengths. Also, $R_{Ai}$ and $R_{Bi}$ represent the impulse responses of spatio-temporal filter; whose spatial part is obtained by differentiating the Gaussian signal $G_m(x,\sigma) = \frac{\partial^m}{\partial x^m}\left\{\frac{1}{2\pi\sigma^2}\exp\left(-x^2 / 2\sigma^2\right)\right\}$, $x \in R$ ($m$ and $\sigma$, respectively, denote the spatial order and scale) and whose temporal part is obtained from the exponential signal $F_n(t) = \left(\frac{1}{n!} - \frac{(kt)^2}{(n+2)!}\right)(kt)^n \exp(-kt)$, $0 \le t \le T$ ($n$ and $k$, respectively, indicate the temporal order and scale). Now in each frame, the gait direction (or gait energy, $EN_L$ and $EN_R$) is easily determined by the convolution of impulse responses in the silhouette image and the summation of directions:

$$\begin{cases} EN_L(t) = R_{L1}^2(t) + R_{L2}^2(t) \\ EN_R(t) = R_{R1}^2(t) + R_{R2}^2(t) \end{cases}, \tag{2}$$

that is, $R_{Li}$ and $R_{Ri}$ indicate the response of filtering obtained from convolution of silhouette with kernels in Eq. (1). The final gait direction in each frame is determined by subtracting the responses. Also, the GSI template is obtained by collecting the responses in every half-period and by averaging the responses of two successive half-periods:

$$GSI = \frac{1}{2}\sum_{k=1}^{2} EN_k, \tag{3}$$

in which, $EN$ represents the sum of the responses in every half-period.

In the GSI approach, because of ignoring human gait in the vertical direction and also using a network of nonlinear filters, the computed responses will be rather unrealistic. So by employing a simple and realistic template in the GSTI method, the researchers have attempted to obtain a more precise human gait template. First, it has been assumed that an input video is a circuit network and that every pixel in it is equivalent to a resistor which is connected to adjacent pixels in the current frame and preceding frame. Also, in the network, the luminosity of each pixel has been considered as the load of a capacitor; and the goal is to calculate current flux in the circuit network. By solving the relevant equations and using Fourier transformation, the spatial impulse response ($S_{IR}$) and the temporal impulse response ($T_{IR}$) for human gait are obtained as follows [9]:

$$S_{IR} = LoG(x,y;\sigma) = \frac{\partial^2}{\partial x^2}G(x,y;\sigma) + \frac{\partial^2}{\partial y^2}G(x,y;\sigma), \tag{4}$$

$$T_{IR} = \frac{1}{2}a_0 + a_1\cos(\omega_\tau(t-\tau)) \quad t \ge 0. \tag{5}$$

In the above equations, $G(.)$ indicates the standard Gaussian signal, $a_0$ and $a_1$ are the coefficients of Fourier series, $\omega_\tau = \pi/T$ ($T$ being the gait period), and $\tau = T/2$. It should be mentioned that in computing Eqs. (4) and (5), some limitations have been applied on Fourier transform solution to make it more compatible with human gait model [9]. Furthermore, the temporal response of Eq. (5) is an even signal; and after computing its corresponding odd signal, the squares of temporal filter responses are added together. After calculating the impulse responses of spatial and temporal filters, the shadow image is first convoluted with a spatial filter and then with a temporal filter in order to obtain the corresponding response. Eventually, similar to the GSI method, the responses in every half-period are collected and the average energy in a period is calculated as the GSTI template.

## 3.3 Local Patch-Based Gait Features

The idea of patch-based approach is different from the previous methods. Here we are trying to seek the important regions in filter responses and to keep them in the final template. Therefore, there are three major steps in the PGF method [11]: (1) filtering the silhouette images and extracting the local patches, (2) calculating the probability distributions of patches, and (3) computing the PGF template. In the first step, the local patches, $p(x,y,t)$, are extracted by seeking the extremum points in a local sliding window in 3-D space. In this step, the response of Gabor filter for each silhouette image is calculated and becomes a basis for seeking the patches. In the second step, the histograms of patch distributions along the $x$, $y$, and $t$ directions are computed separately:

$$\begin{aligned} h_x &= hist\left[p(x_i,y,t)\right]_{i=1\ldots H} \\ h_y &= hist\left[p(x,y_i,t)\right]_{i=1\ldots W}, \\ h_t &= hist\left[p(x,y,t_i)\right]_{i=1\ldots T} \end{aligned} \tag{6}$$

Here, $H$, $W$, and $T$ are the image height and width and the gait period, respectively. The above triple histogram is unique to each individual and can be used as gait signature [11].

Now in the third step, by considering the above probability distributions as coordinate weights, the final template can be obtained. Here, the set of 40D Gabor responses (with five different scales and eight directions) is applied on the GEI, and the weighted coordinates of each pixel are added to the responses.

$$PGF = \left[RG^T, h_i k(X_i), h_j k(Y_j), h_k k(t_k)\right]^T_{i=1\ldots H, j=1\ldots W, k=1\ldots T}, \tag{7}$$

in which, $RG$ is response of the 40D Gabor filtering, $k(.)$ is the isotropic kernel for allocating more weight to the pixels near the image center or half-period, and $h_l$ is the weight resulting from the relevant histogram index. According to Eq. (7), the PGF template is an augmented Gabor template in which the spatio-temporal coordinates of patches have been added to filter response based on their probability distribution.

### *3.4 Classification*

The introduced methods highlighted the manner of extracting powerful features from behavior template. However, the usefulness of recent approaches depends not only on feature extraction but also on the type of classification. The simplest method of classifying a template is to find the closest feature to a gallery determined by direct template matching. For example, the 1-nearest neighbor classification technique (1-NN) assigns a label to its nearest neighbor (with the least Euclidean distance) [8]. However, since feature dimensions are much more than the number of test samples and some features are overlooked, we have an undersample problem (UPS), which diminishes the recognition accuracy [6]. A common method for dealing with this problem is to use the two-step technique of PCA+LDA [10]. This technique is relatively simple and provides usual accuracy; because by applying the PCA matrix, the 2-D structure of data collapses and data turn into a one-dimensional vector. Despite the existence of various challenges, the random subspace method (RSM) has been able to identify gait templates with high accuracy [11, 27]. This classification method achieves weak classification by taking random samples from feature space and by making a decision based on a set of weak decisions. Considering the utility of the RMS classification technique, it has been employed in this paper. More details on this classification scheme can be found in the work of Guan et al. [27].

## 4 System-Level Evaluation Tools

So far, we explored the algorithms used for identifying behavior templates. However, quantitative measures are needed to evaluate the performances of these algorithms. By means of these quantitative criteria we can determine the utility and fidelity of different identification approaches. Every biometric system is evaluated based on different parameters so as to obtain its fidelity under different conditions. The existing approaches for evaluating the fidelity of biometric systems have been developed based on the performance of face and fingerprint biometrics [25, 28]; however, this procedure can also be extended to a gait biometric system.

In behavioral biometry systems the results are usually reported in terms of identification accuracy [28]; because the number of test sets is limited and the individuals in probe set are always a subset of the individuals in gallery. Moreover, one of the main challenges of behavioral biometry systems is the decline of their performance in uncontrolled or unpredictable conditions [28]. Basically, the evaluation of every biometric system requires three parameters [28]: (1) evaluating the quality of input data (sensor quality), (2) guaranteeing the performance of algorithm, and (3) providing a utility benchmark. In this paper, the quality of input data is the same for all the evaluated algorithms; because the same standard datasets have been used for all of them. Also, the utility of the proposed algorithms in properly identifying gait templates has been confirmed. Therefore, the above three criteria [25], which are intra-system evaluation measures, are not sufficient for evaluating the performance of a gait biometric systems.

We define three important criteria for the extra-system evaluation of biometry systems: (1) system error, (2) execution speed and the amount of computer memory used, and (3) scalability. Each of these criteria will be subsequently explained.

## 4.1 *False Match Rate (FMR) and False Non-match Rate (FNMR) Criteria*

These two benchmarks are able to compute the amount of error for a system. The FMR criterion is an empirical estimation of probability (percent occurrence), by which a system falsely accepts an input template. Also, FNMR is an empirical estimation of probability, by which a system falsely rejects a matched input sample. FMR and FNMR criteria are equivalent to FAR and FRR, respectively; and for their computation procedures one can consult the available references [25]. Ordinarily, when comparing two biometry systems, the system which has a lower FRR at the same FAR is the better performing system.

The recognition error tradeoff (DET) graph has been used to graphically display system error; because it is not commonplace to use the ROC diagram in the evaluation of biometry systems [29]. The horizontal and the vertical axes of the DET graph, respectively, show the FAR and FRR values at different accuracies and Ranks. In DET diagram standards, the vertical axis is in percentage (%) and the horizontal axis has a logarithmic scale.

## 4.2 *Computational and Memory Complexity*

The time needed for identifying an individual and the amount of computer memory used for this purpose are determined by this criterion. In fact, time complexity and used memory constitute suitable measures for evaluating a system. When comparing two systems with the same error, the one with higher speed and less memory usage is considered a better system. A biometric system needs memory to compute the templates and to save the features.

### 4.3 Scalability

This criterion expresses the performance of a system in huge set of data. To ensure a level of scalability that can support a very large dataset, it is necessary to use very strong support servers of high capacity. Assuming that the processing power of each server is one unit, the total processing power of ten servers with integrated processing capability is not 10 units. Even by increasing hardware and software resources, there will be a problem with system performance at large volumes of data; because with the rise in the number of samples, the degree of differentiability between samples decreases severely and system error goes up. In other words, a system's robustness against scalability indicates the reliability of that system under real conditions. Obviously, in comparing two systems with equal errors, the one with a lower error in a high-population dataset is a better system. Published papers do not present a specific criterion for assessing a system's robustness to scalability. In this paper, average Rank-1/Rank-5 accuracy achieved in three datasets will be used to evaluate the scalability criterion.

## 5 Results and Discussion

In this section, the results of algorithm- and system-level evaluations will be presented. Algorithm-level evaluation has been thoroughly discussed in the cited references [9–11]. In this paper, the results of algorithm-level evaluations are concisely presented and then the results of system-level evaluations are discussed. An algorithm-level comparison is carried out in order to assess the quality of the examined algorithms in gait template recognition; because an exact biometric system is dependent on the type of algorithm designed for it. The algorithms that have been selected for evaluation and comparison are as follows: GEI [1], CGI [8], 5/3GI [3], STIP [30], locality-constrained group sparse representation (LGSR) [20], view-invariant multiscale gait recognition (VI-MGR) [2], RSM [27], local patch-based subspace ensemble learning algorithm (LPSELA) [4], two-stream generative adversarial network (TS-GAN) [31], clothing-invariant gait recognition (CLIGR) [15], and local feature at the bottom layer based on capsule network (LBC) [18].

In this paper, three famous datasets (USF [12], CASIA-B [13], and OU-ISIR-B [14]) are used for algorithm evaluation. The USF dataset contains 122 individuals with five walking forms, CASIA-B has 124 individuals with ten walking forms, and the OU-ISIR-B dataset includes 48 individuals with 32 different forms of walking. With different test cases in the gallery and the probe sets, there are a total of 3176 unique tests for these three datasets, (1080 tests for USF, 1240 tests for CASIA-B, and 856 tests for OU-ISIR-B). For evaluating the examined algorithms, each of the above datasets has considered several gait challenges. The USF set has addressed the variations of walking manner with the change of shoes, viewing conditions, surface, bag, and time parameters. The CASIA-B set has considered gait variations with respect to "normal," "bag," and "cloth" conditions at different video recording angles. And the OU-ISIR-B set has only dealt with the change of clothing, which is the most important and common parameter related to gait. In the following, the results of algorithm-level evaluations on the three mentioned datasets will be briefly discussed.

## 5.1 Algorithm-Level Evaluations

To verify the effective performance of the proposed algorithms, the first test is carried out on the CASIA-B dataset. Considering the high quality of silhouette images in this dataset, there is no need to use powerful classification techniques such as RSM; and satisfactory accuracy will be obtained by applying 1-NN. For a brief and accurate evaluation of the examined approaches, only Rank-1 values are presented. Table 1 shows the performance Rank-1 values achieved by the proposed algorithms.

Compared to the first group of algorithms, the average accuracy of Rank-1 is higher than that of GEI, CGI, and 5/3GI methods. The average accuracy of GSTI is close to that of CGI, which shows an acceptable performance. Also, in five tests (out of 6), the proposed approaches achieve a higher performance relative to GEI and CGI. The results of 5/3GI for certain experiments have not been revealed and their averages cannot be evaluated.

In the second part of Table 1, the satisfactory performance of the proposed algorithms can be revealed by comparing the results through multiview-based methods. Compared to STIP algorithm, the proposed algorithms are superior in three tests out of six; and in some other tests the results are very close. In comparison with the VI-MGR method, in two cases our algorithms have achieved greater performance. However, because of using synthetic information and relatively powerful classification technique, the VI-MGR algorithm has attained a higher accuracy. In summary, in the CASIA-B dataset, the proposed gait identification algorithms have achieved satisfactory or better recognition accuracies than other recent approaches.

In the second step, the performances of the proposed gait recognition algorithms in the OU-ISIR-B dataset are investigated. Here, the Rank-1 results obtained by

**Table 1** Evaluating the Rank-1 recognition rates achieved by the proposed algorithms and other approaches in CASIA-B dataset

| Gall-Prb | nm-nm | nm-bg | nm-cl | bg-bg | bg-cl | cl-cl | Avg. |
|---|---|---|---|---|---|---|---|
| GEI [1] | 91.57 | 37.8 | 25.04 | 91.2 | 17.47 | 97.22 | 60.05 |
| CGI [8] | 88.06 | 51.93 | 46.88 | 89.81 | 32.52 | 95.37 | 67.43 |
| 5/3GI [3] | 98 | – | – | 73 | 66 | – | – |
| STIP [30] | 95.4 | 60.9 | 52 | 73 | 29.8 | 70.6 | 63.62 |
| VI-MGR [2] | 100 | 89 | 76 | 79.03 | – | 71.77 | – |
| TS-GAN [31] | – | – | – | – | – | – | 63.1 |
| GSI [10] | 95.87 | 66.46 | 41.41 | 90.5 | 29.5 | 100 | 70.62 |
| GSTI [9] | 91.8 | 57.5 | 39.6 | 92.5 | 26.4 | 98 | 67.64 |
| PGF [11] | 92.4 | 65.7 | 50.8 | 91.5 | 33.5 | 97.5 | 71.9 |

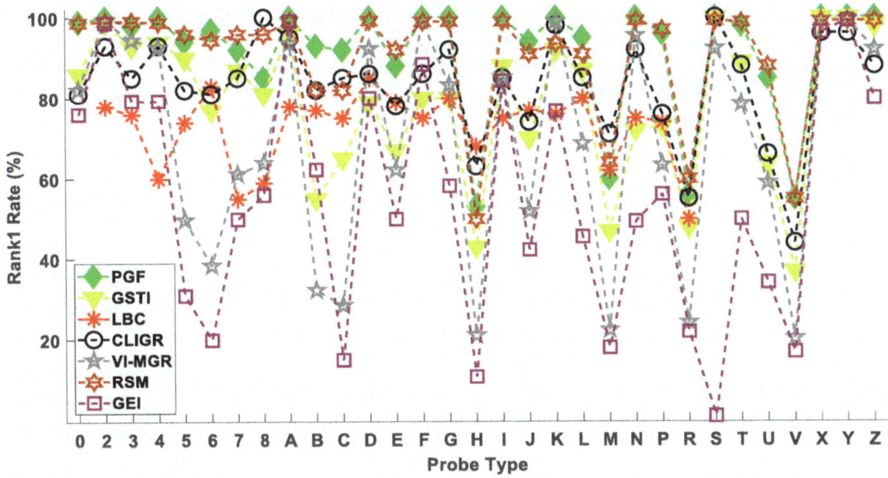


**Fig. 2** Evaluating the performances of various algorithms in the OU-ISIR-B dataset

**Table 2** Comparing the average Rank-1 results obtained by the proposed and more recent algorithms in the OU-ISIR-B dataset

| Algorithms | Avg. of Rank-1 (%) |
|---|---|
| GEI [1] | 55.6 |
| RSM [27] | 90.7 |
| VI-MGR [2] | 69.03 |
| CLIGR [15] | 81.71 |
| LBC [18] | 73.33 |
| GSTI [9] | 77.25 |
| PGF [11] | 90.62 |

these algorithms in the probe set (which includes 31 different combinations of clothing) have been presented. However, due to the proximity of the results of GSI and GSTI algorithms, the GSI results have not been listed. Figure 2 graphically displays the fidelities of the examined algorithms when considering different clothing conditions. Also, Table 2 shows the average accuracies achieved by the proposed algorithms in the OU-ISIR-B dataset.

According to Fig. 2, in five tests (probes D, F, G, K, and N), the VI-MGR algorithm achieves a higher accuracy than GSTI. Also, in comparison with RSM, the results of GSTI can still be observed. Compared to the RSM approach, the GSTI algorithm uses a simpler filter structure; because instead of the 40-component Gabor functions, it exploits spatial and temporal impulse responses.

In addition, the PGF algorithm (which has a patch-based structure) enjoys a higher accuracy and fidelity relative to the GSTI algorithm. More precisely, in five tests out of 31 (probes 2, S, X, Y, and Z), the recognition accuracy of PGF is 100% and the algorithm has been able to identify all the individuals in the probe set. Also in 2 probes out of 31 (probes H and M), the patch-based methods exhibit weaker

**Table 3** Rank-1 and Rank-5 results of different algorithms obtained for the USF dataset

| Exp. | A | B | C | D | E | F | G | H | I | J | K | L | Avg. |
|---|---|---|---|---|---|---|---|---|---|---|---|---|---|
| Method | Rank-1 performance | | | | | | | | | | | | |
| LGSR | 95 | 93 | 89 | 51 | 50 | 29 | 36 | 85 | 83 | 68 | 18 | 24 | 70.07 |
| RSM | 100 | 95 | 94 | 73 | 73 | 55 | 54 | 97 | 99 | 94 | 41 | 42 | 81.15 |
| VI-MGR | 95 | 96 | 86 | 54 | 57 | 34 | 36 | 91 | 90 | 78 | 31 | 28 | 68.13 |
| LPSELA | 95 | 91 | 78 | 66 | 59 | 46 | 52 | 93 | 88 | 69 | 30 | 27 | 70.49 |
| GSI | 92 | 95 | 86 | 38 | 31 | 17 | 24 | 79 | 83 | 76 | 25 | 19 | 58.44 |
| GSTI | 97 | 95 | 93 | 53 | 49 | 41 | 46 | 96 | 97 | 92 | 33 | 21 | 72.25 |
| PGF | 100 | 96 | 98 | 62 | 59 | 43 | 46 | 100 | 99 | 94 | 28 | 30 | 76.01 |
| | Rank-5 performance | | | | | | | | | | | | |
| LGSR | 99 | 94 | 96 | 89 | 91 | 64 | 64 | 99 | 98 | 92 | 39 | 45 | 85.31 |
| RSM | 100 | 98 | 98 | 85 | 84 | 79 | 73 | 98 | 99 | 98 | 55 | 58 | 88.59 |
| VI-MGR | 100 | 98 | 96 | 80 | 79 | 66 | 65 | 97 | 95 | 89 | 50 | 48 | 83.75 |
| LPSELA | 100 | 96 | 93 | 84 | 83 | 73 | 74 | 95 | 96 | 89 | 64 | 52 | 86.09 |
| GSI | 98 | 95 | 95 | 67 | 52 | 43 | 56 | 97 | 95 | 95 | 43 | 37 | 76.60 |
| GSTI | 100 | 96 | 97 | 78 | 76 | 72 | 74 | 99 | 99 | 99 | 42 | 36 | 85.64 |
| PGF | 100 | 98 | 98 | 80 | 77 | 77 | 60 | 100 | 100 | 99 | 48 | 45 | 86.59 |

performance relative to more recent approaches (CLIGR [15] and LBC [18]). Nevertheless, the performance of other methods under such conditions is very low. The results in the above figure indicate a performance below 60% for the PGF algorithm in probes H, M, R, and V. This is mainly due to the coverage of body parts during patch extraction. In these conditions, the proposed approaches cannot extract enough information from lower body parts. The results of Fig. 2 and Table 2 show the relative superiority of the proposed algorithms compared to the other algorithms.

In the third step of this section, the complete Rank-1 and Rank-5 results of different algorithms obtained for the USF dataset have been discussed.

According to Table 3, the average Rank-1 and Rank-5 results in the GSTI and PGF approaches have improved relative to all other methods (except RSM). Also, the results of PGF are very close to those of the RSM algorithm. The proposed methods are vulnerable to the change of "surface" and "time" (probes F, G, K, and L), but their performance is still comparable to all the other approaches. The Rank-5 recognition rates of the proposed algorithms have improved in most of the tests. More precisely, in three tests out of 12 (probes A, H, and I), the recognition accuracy of PGF is 100% and all the relevant subjects have been identified. Also, in the rest of the tests, the Rank-5 values are close to their maximum, and subjects can be identified with a high accuracy. The obtained results indicate that the average recognition success of PGF has increased by about 5% relative to other methods (e.g., LGSR [20], VI-MGR [2], and LPSELA [4]).

So the proposed algorithms are appropriate tools for human gait identification and provide a suitable basis for recognition of individuals under challenging and more complicated conditions.

## 5.2 System-Level Evaluations

In the previous section, the proposed algorithms for the recognition of behavioral templates were reviewed. However, quantitative measures are needed to evaluate the system-level performance of these algorithms. The utility and the fidelity of the proposed biometric systems can be obtained by means of these quantitative criteria. As it was stated in Sect. 4, the three criteria of system error, algorithm execution speed and the amount of required computer memory, and scalability have been used in this paper. The amount of system error is computed by using the FAR and FRR error parameters and evaluated by means of the detection error tradeoff (DET) diagram. In biometric systems for face recognition, (normally in the authentication mode), the above errors [33] are expressed as "FRR 1% @ FAR 1/10,000". However, since behavioral detection systems act in the identification mode, error calculation will be a bit different. Suppose the $m^{th}$ individual from a probe set is compared with all the individuals of a gallery (G individuals) and that these individuals are arranged descending. Now if the true (or corresponding) individual is located in the $i^{th}$ position, then in computing the Rank-r parameter, the values of FAR and FRR will be calculated as follows:

$$\begin{cases} \text{FAR} = \dfrac{i-1}{G}, & \text{FRR} = \dfrac{r-i}{G}. \quad if\ i \le r \\ \text{FAR} = \dfrac{r}{G}, & \text{FRR} = \dfrac{i-r}{G}. \quad if\ i > r \end{cases} \tag{8}$$

Now by changing the value of r and considering the values of Rank-r and by recalculating the error values according to the above equation, the DET graph is obtained. In the DET diagram standards, the vertical axis shows the FRR percentage and the horizontal axis has a logarithmic scale. Due to its popularity and comprehensiveness, the USF dataset has been used here for error estimation; and the obtained error values are the weighted average errors in the whole probe sets and in different tests. Table 4 shows the weighted average error values with respect to Rank-1 and Rank-5 in the proposed systems. Also, Fig. 3 illustrates the DET diagrams for these systems in the USF dataset.

According to Table 4, the PGF system has the least values of FRR and FAR; so it can be considered as a powerful and low-error system. However, a proper comparison cannot be made based on the values in the above table; because in comparing two biometric systems, the better system will have a lower FRR at the same FAR

**Table 4** The specifications of the proposed biometric systems versus Rank-1 and Rank-5 errors

| Biometric system | @Rank-1 | @Rank-5 |
|---|---|---|
| GSI | FRR 33% @ FAR 49/1000 | FRR 16% @ FAR 51/1000 |
| GSTI | FRR 24% @ FAR 48/1000 | FRR 10% @ FAR 49/1000 |
| PGF | FRR 23% @ FAR 38/1000 | FRR 9% @ FAR 40/1000 |

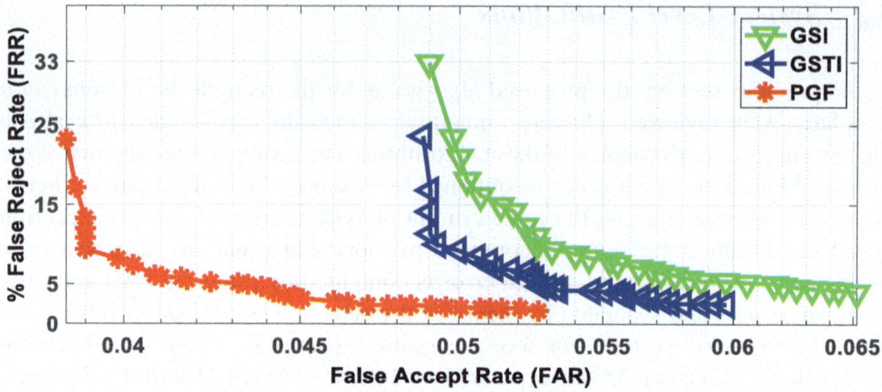


**Fig. 3** The DET diagrams resulting from the three proposed systems

value. For a more appropriate evaluation, the DET diagram has been used. As Fig. 3 shows, the PGF system has a much lower error than the other two systems. At a specific FAR (FAR = 10–1.3 ≈ 50/1000), the FRR value of PGF (2%) is less than that of GSTI (10%), which itself is less than that of GSI (18%). A 16% difference exists between the FRR values of PGF and GSI systems. So the PGF system has a lower gait recognition error.

Now the systems are evaluated in terms of processing time and required computer memory. The most important part of the proposed systems is their input image filter. The GSI system uses a quadruple directional filter (Eqs. (1) and (2)), the GSTI system uses a dual spatio-temporal filter (Eqs. (4) and (5)), and the PGF system employs a 40-component Gabor filter (Eq. (7)). Now if the image dimensions are *WxH* and the kernel dimensions are *wxh*, the approximate computation time of image filter will be of order $O(I_{filt}) \approx O(WHwh)$ and the memory used will be of order $O(I_{filt}) \approx O(WH + wh)$ [10]. Therefore, the time complexities of GSI, GSTI, and PGF systems in a dataset with the number of training and test individuals equal to $n_{te} + n_{tr}$ and with a gait period of $T$ will, respectively, be $O(4(n_{te} + n_{tr})TWHwh)$, $O((n_{te} + n_{tr})TWHwh)$, and $O(10(T + 1)(n_{tr} + n_{te})WHwh)$ [9–11]. Also, by assuming $WH >> wh$, the amounts of memory required by these systems are $O(10TWH(n_{te}+n_{tr}))$, $O(3TWH(n_{te} + n_{tr}))$, and $O(10(T + 1)WH(n_{te} + n_{tr}))$, respectively [9–11]. Obviously, time complexity and required memory depend on three main parameters: (1) the size of input image (*WxH*), (2) the total number of individuals in the gallery and probe sets ($n = n_{tr} + n_{te}$), and (3) the average gait period ($T$). It should be mentioned that the dimensions of filter kernel in the examined systems have been considered as $w \times h = 39 \times 39 = 1521$ [9]. If the same dimensions are considered for input image ($W = H = N$), then the required execution time, $t$, for the proposed systems in the whole dataset will be as;

$$\begin{aligned} &\text{GSI}: \quad t \approx O\left(4 \times 1521 \times nTN^2\right) \\ &\text{GSTI}: \quad t \approx O\left(1521 \times nTN^2\right) \\ &\text{PGF}: \quad t \approx O\left(10 \times 1521 \times (T+1) nN^2\right). \end{aligned} \tag{9}$$

Also, the amounts of memory ($M$) used by the proposed systems are as follows:

$$\begin{aligned} &\text{GSI}: \quad M \approx O\left(10 \times nTN^2\right) \\ &\text{GSTI}: \quad M \approx O\left(3 \times nTN^2\right) \\ &\text{PGF}: \quad M \approx O\left(10 \times (T+1) nN^2\right). \end{aligned} \tag{10}$$

Suppose the number of individuals in a dataset (considering the three evaluated datasets) has three default values of $n = \{150, 500, 1000\}$, which correspond to small, medium, and large datasets, respectively; and suppose the gait period has two default values of $T = \{30, 70\}$, which, respectively, specify small and large sequences of recorded human gait. With these assumptions, Figs. 4 and 5 illustrate the systems' computational loads and required memories versus the dimensions of input images and in terms of predefined parameters.

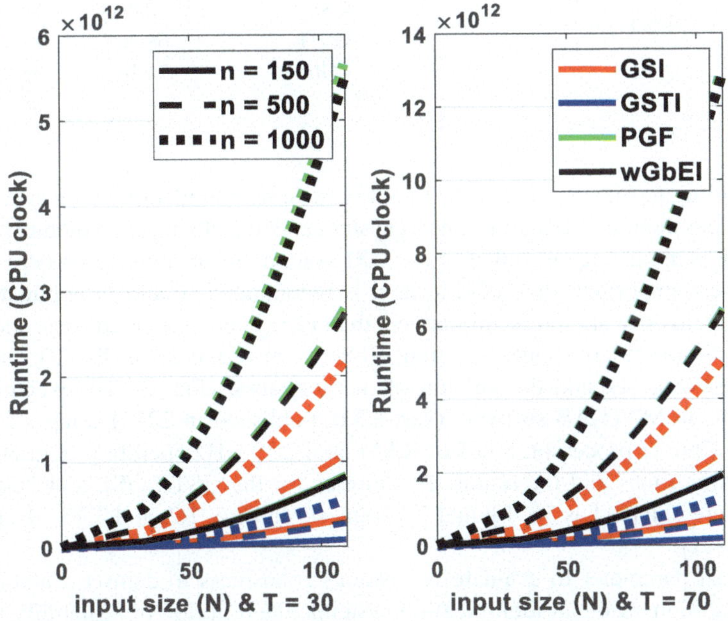


**Fig. 4** Computation times required by the proposed systems versus image dimensions

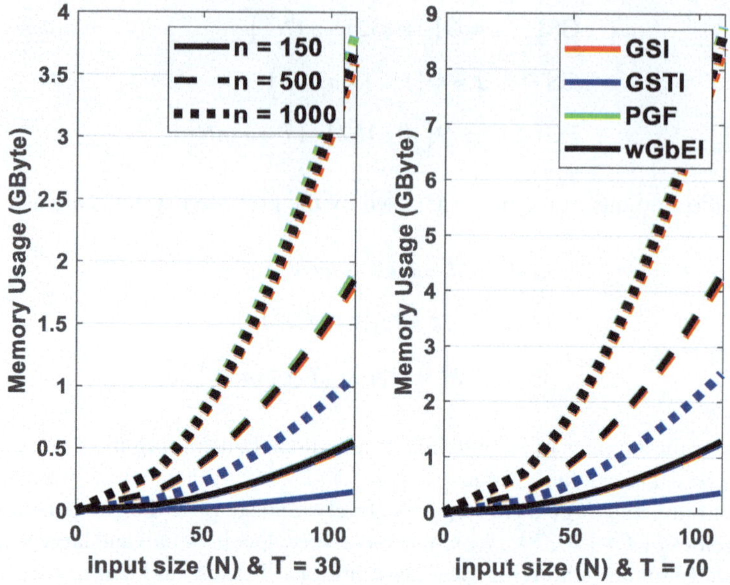


**Fig. 5** The amounts of memory (in *GB*) required by the proposed systems at various settings

**Table 5** Average system accuracies for all the tested datasets (3176 data in total)

| Biometric system | Rank-1 (%) | Rank-5 (%) |
|---|---|---|
| GSI | 70.89 | 84.32 |
| GSTI | 68.96 | 82.47 |
| PGF | 74.16 | 85.28 |

According to Figs. 4 and 5, GSTI is the fastest human identification system that requires less memory. Also, the slowest system is PGF, which uses a memory capacity which is almost equal to that of the GSI system. So in situations where system error is less important, the GSTI system is recommended; and in applications in which system error should be minimized, the PGF system will be the best choice. In these conditions, there should be a tradeoff between system error (Fig. 3), computation speed (Fig. 4), and the amount of memory used (Fig. 5). By executing the algorithms in MATLAB software (ver. 8.3.0, published in 2014) using a PC with Intel (R) Core i7 processor, 8 GB of RAM, and 2.39 GHz operating frequency, the computation times and the required memories for the GSTI, GSI, and PGF algorithms were obtained as 56, 14, and 5.5 frames/s and 3.6, 1.2, and 3.75 GB, respectively [9–11].

The last parameter to evaluate is system's robustness to dataset scalability. In evaluating biometric gait identification systems, the criterion of scalability has not been usually discussed [5]. So in order to measure the degree of robustness against scalability, the average results of the examined systems for all the tested datasets are

**Table 6** Evaluating the fidelity of the proposed biometric systems

| System benchmark | Superior system | Second superior system | Weakest system |
|---|---|---|---|
| FAR & FRR | PGF | GSTI | GSI |
| Time/memory | GSTI | GSI | PGF |
| Scalability | PGF | GSI | GSTI |

computed in this section. A total of 3176 tests have been carried out (1080 tests on the USF dataset, 1240 tests on the CASIA-B dataset, and 856 tests on the OU-ISIR-B dataset). The average results have been tabulated in Table 5.

According to Table 5, the PGF system achieves a higher Rank-1/Rank-5 accuracy and performs better when a large dataset is involved. Also, the GSTI approach proves to be the weakest in this case. So the PGF system will have a better performance in a large-volume dataset.

### *5.3 System-Level Performance Evaluation Criteria*

In the preceding section, the proposed systems were thoroughly evaluated. From the perspective of FAR and FRR errors, a system may produce minimal error but have a large computational load and require extensive memory. Also, error optimization is equivalent to system's robustness against the increase of dataset volume, and vice versa. In evaluating the proposed systems, Table 6 ranks the proposed biometric gait identification systems.

According to the above table, if the amount of computer memory used and the computation time are not important factors, the PGF system will be the most suitable, because of its lower error. Also, when the execution speed of algorithm and the memory it uses are of utmost importance, the GSTI system is recommended. In this case, the GSI system can also be used as a substitute system for GSTI; both of these systems have a greater computation speed and require less memory compared to the PGF system.

## 6 Conclusion

In this paper, effective gait template detection algorithms were reviewed and the evaluation criteria for the biometric systems related to these algorithms were presented. Today, there are three major challenges in properly identifying gait templates: (1) using temporal information, (2) making the algorithms compatible with human gait templates, and (3) eliminating extra information and noise from final features. As was previously mentioned, the three algorithms of GSI, GSTI, and PGF have been able to deal with the above three problems and to improve the fidelity of gait template recognition in most cases. However, for a more exact evaluation of performance, these algorithms have been compared from the standpoint of biometric systems. For this purpose, three criteria of system error, computational load, and scalability have been defined and used to evaluate system performances. The fastest and the most accurate biometric systems are GSTI and PGF, respectively. But when the computational load is limited and a relatively low system error is expected, the GSI system could be reliably used as an alternative system. In general, the criteria presented in this paper can be used as a standard measure to evaluate biometric gait identification systems and to help us choose the right system suitable for real conditions.

## References

1. Han, J., & Bhanu, B. (2006). Individual recognition using gait energy image. *IEEE Transactions on Pattern Analysis and Machine Intelligence, 28*(2), 316–322.
2. Choudhury, S. D., & Tjahjadi, T. (2015). Robust view-invariant multiscale gait recognition. *Journal of Pattern Recognition, 48*(3), 798–811.
3. Atta, R., Shaheen, S., & Ghanbari, M. (2017). Human identification based on temporal lifting using 5/3 wavelet filters and radon transform. *Journal of Pattern Recognition, 69*, 213–224.
4. Ma, G., Wua, L., & Wang, Y. (2017). A general subspace ensemble learning framework via totally-corrective boosting and tensor-based and local patch-based extensions for gait recognition. *Pattern Recognition, 66*, 280–294.
5. Rida, I., Almaadeed, N., & Almaadeed, S. (2019). Robust gait recognition: A comprehensive survey. *IET Biometrics, 8*(1), 14–28.
6. Tao, D., Li, X., Wu, X., & Maybank, S. J. (2007). General tensor discriminant analysis and Gabor features for gait recognition. *IEEE Transactions on Pattern Analysis and Machine Intelligence, 29*(10), 1700–1715.
7. Lam, T. H., Cheung, K. H., & Liu, J. N. (2011). Gait flow image: A silhouette-based gait representation for human identification. *Journal of Pattern Recognition, 44*(4), 973–987.
8. Wang, C., Zhang, J., Wang, L., Pu, J., & Yuan, X. (2012). Human identification using temporal information preserving gait template. *IEEE Transactions on Pattern Analysis and Machine Intelligence, 34*(11), 2164–2176.
9. Ghaeminia, M. H., & Shokouhi, S. B. (Feb. 2019). On the selection of spatio-temporal filtering with classifier ensemble method for effective gait recognition. *Journal of Signal, Image and Video Processing, 13*(1), 43–51.
10. Ghaeminia, M. H., & Shokouhi, S. B. (2018). GSI: efficient spatio-temporal template for human gait recognition. *International Journal of Biometrics (IJBM), 10*(1), 29–51.
11. Ghaeminia, M. H., Shokouhi, S. B., & Badiezadeh, A. (2019). A new spatiotemporal patch-based feature template for effective gait recognition. *Multimedia Tools and Applications (MTAP), 79*(1-2), 713–736. https://doi.org/10.1007/s11042-019-08106-x.
12. Sarkar, S., Phillips, P. J., Liu, Z., Vega, I. R., Grother, P., & Bowyer, K. W. (2005). The humanID gait challenge problem; data sets, performance, and analysis. *IEEE Transactions on Pattern Analysis and Machine Intelligence, 27*(2), 162–177.
13. Yu, S., Tan, D., & Tan, T. (2006). A framework for evaluating the effect of view angle, clothing and carrying condition on gait recognition. In *Proceeding of 18th International Conference on Pattern Recognition (ICPR)*. Hong Kong: IEEE Computer Society.
14. Iwama, H., Okumura, M., Makihara, Y., & Yagi, Y. (2012). The OU-ISIR gait dataset comprising the large population dataset and performance evaluation of gait recognition. *IEEE Transactions on Information Forensics and Security, 7*(5), 1511–1521.

15. Ghebleh, A., & Moghaddam, M. E. (2018). Clothing-invariant human gait recognition using an adaptive outlier detection method. *Journal of Multimedia Tools and Applications (MTAP), 77*(7), 8237–8257.
16. Hong, S., Lee, H., & Kim, E. (2013). Probabilistic gait modelling and recognition. *IET Computer Vision, 7*(1), 56–70.
17. Ben, X., Gong, C., Zhang, P., Jia, X., & Meng, Q. W. (2019). Coupled patch alignment for matching cross-view gaits. *IEEE Transactions on Image Processing, 28*(6), 3142–3157.
18. Xu, Z., Lu, W., Zhang, Q., Yeung, Y., & Chen, X. (Feb. 2019). Gait recognition based on capsule network. *Journal of Visual Communications and Image Representation, 59*, 159–167.
19. Huang, Y., Xu, D., & Nie, F. (2012). Patch distribution compatible semisupervised dimension reduction for face and human gait recognition. *IEEE Transactions on Circuits and Systems for Video Technology, 22*(3), 479–488.
20. Xu, D., Huang, Y., Zeng, Z., & Xu, X. (2012). Human gait recognition using patch distribution feature and locality-constrained group sparse representation. *IEEE Transactions on Image Processing, 21*(1), 316–326.
21. Fendri, E., Chtourou, I., & Hammami, M. (2019). Gait-based person re-identification under covariate factors. *Journal of Pattern Analysis and Application (PAA), 22*(4), 1629–1642. https://doi.org/10.1007/s10044-019-00793-4.
22. Tong, S., Fu, Y., & Ling, X. Y. (2018). Multi-view gait recognition based on a spatial-temporal deep neural network. *IEEE Access, 6*, 57583–57596.
23. Grother, P., & Tabassi, E. (April 2007). Performance of biometric quality measures. *IEEE Transactions on Pattern Analysis and Machine Intelligence, 29*(4), 531–543.
24. Teixeira, R. F., & Leite, N. J. (2017). A new framework for quality assessment of high-resolution fingerprint images. *IEEE Transactions on Pattern Analysis and Machine Intelligence, 39*(10), 1905–1917.
25. Alonso-Fernandez, F., Fierrez, J., & Ortega-Garcia, J. (2012). Quality measures in biometric systems. *IEEE Security & Privacy, 10*(6), 52–62.
26. Olsen, M. A., Smida, M., & Busch, C. (2016). Finger image quality assessment features – Definitions and evaluation. *IET Biometrics, 5*(2), 47–64.
27. Guan, Y., Li, C.-T., & Roli, F. (2015). On reducing the effect of covariate factors in gait recognition: A classifier ensemble method. *IEEE Transactions on Pattern Analysis and Machine Intelligence, 37*(7), 1521–1528.
28. Ward, J. A., Lukowicz, P., & Gellersen, H. (2011). Performance metrics for activity recognition. *ACM Transactions on Intelligent Systems and Technology, 2*(1), 1.
29. Bolle, R. M., Connell, J. H., Pankanti, S., & Senior, N. K. (2005). The relation between the ROC curve and the CMC. In *Fourth IEEE workshop on automatic identification advanced technologies (AutoID'05). Buffalo, NY, USA*.
30. Kusakunniran, W. (2014). Recognizing gaits on spatio-temporal feature domain. *IEEE Transactions on Information Forensics and Security, 9*(9), 1416–1423.
31. Wang, Y., Song, C., Huang, Y., Wang, Z., & Wang, L. (2019). Learning view invariant gait features with two-stream GAN. *Journal of Neurocomputing, 33*, 245–254.

# Interactive Biometric Identification System Based on the Keystroke Dynamic

**Stepan Bilan, Mykola Bilan, and Andrii Bilan**

## 1 Introduction

In today's world, the world's population is constantly increasing. Along with the population, the number of computing tools and automation tools that accompany people is also increasing. Each person on earth has unique individual characteristics that are inherent in him alone. Such characteristics of a person can be divided into static and dynamic. Static characteristics include fingerprint image, face image, handwritten image, ear shape, hand shape, finger vein pattern, retina image, iris image, etc. The dynamic characteristics include gait, the dynamics of writing handwriting, the dynamics of the handwriting of the keyboard, the movement of the lips, the dynamics of the heart, etc.

Population growth requires solving such a problem as automatic biometric identification of a person. This task is especially faced in access systems, in systems for the interaction of people over long distances, in video surveillance systems, in transport and other areas. To solve the problems of biometric identification, a variety of intelligent methods of information processing are used. For each biometric characteristic, a separate method is developed and implemented, on the basis of which a biometric identification system is developed. For static biometric characteristics, image processing methods and optoelectronic signal conversion tools are mostly used. For dynamic biometric characteristics, methods of processing the obtained ordered data set are used (for example, a sequence of video data).

The use of statistical and dynamic biometric characteristics entails certain advantages and disadvantages. In particular, statistical biometric characteristics have low

S. Bilan (✉)
State University of Infrastructure and Technology, Kyiv, Ukraine
e-mail: bilan_sm@gsuite.duit.edu.ua; nsuhov@idknet.com

M. Bilan · A. Bilan
The Municipal Educational Institution Mayakskaya Secondary School, Mayak, Moldova



S. Bilan et al. (eds.), *Biometric Identification Technologies Based on Modern Data Mining Methods*, https://doi.org/10.1007/978-3-030-48378-4_3

resistance to various enemy attacks. Statistical biometric characterization is not difficult to fake and use it for a long time. This situation limits the use of statistical biometric characteristics.

In terms of resistance to enemy attacks, dynamic biometric characteristics have more advantages than static. They are almost impossible to recreate by another person or automated tool. However, identification systems based on biometric dynamic characteristics occupy a small percentage of the world market and are currently being developed and rapidly implemented.

## 2 Formulation of the Problem

Currently, computer technology is widely used, which requires data input using the keyboard [1–4]. Especially important are the tasks of determining the identity of the person entering this data. Such data may be text, access key, etc. People are less and less using a regular pen, and more often they are using a keyboard and electronic notebooks. Currently, the keyboard is increasingly replacing the conventional pen for writing on paper. A large number of biometric identification systems for the dynamics of the keyboard have already been developed. However, due to the fact that they can be applied to specialists who have been working with a keyboard for a long time, such systems are still not widely used. This is also due to the various instabilities of this biometric characteristic, since it can change at different times of the day or year, and also depends heavily on the person's internal emotional state. The purpose of this chapter is to increase the accuracy of biometric identification through the use of a fixed and free key sequence, as well as through the use of additional characteristic features. The problem is also solved by selecting the necessary threshold values for text sequences of different lengths.

## 3 The Existing Works in This Area

Today, the use of biometric methods for access control is receiving a lot of attention. Keyboard handwriting is a set of dynamic parameters that appear when working on the keyboard. The user's personality is the speed of typing, various habits of keystrokes, etc. Such features and habits allowed us to create a number of methods for biometric identification by the dynamics of the handwriting of the keyboard [5–10].

The classical statistical approach to keyboard handwriting showed a number of features [6, 8, 10, 11]:

- dependence of handwriting on the combination of characters in the word;
- deep connections between a set of different characters;
- delays when entering characters.

An important parameter of such identification is the presence of a passphrase, which consists of a sequence of characters placed on the keyboard.

In modern systems, user authentication is based on a passphrase. This turns on the dynamics of keystrokes. In this case, the knowledge of the password and the method of entering it are taken into account. Such dynamics and authentication process are already hacked [3].

In modern systems, authentication is used during typing of any text in a session [11]. Such authentication is called continuous [3]. If another person has been working in the system for a long time, the system can determine the substitution.

For successful authentication using keyboard handwriting, distance-based studies [12], statistical properties [13], and bioinformatics tools are used [14]. These studies were conducted to evaluate the received data from one user. If data from a true user and from an impostor were used, then authentication methods were built on neural networks [15] and on support vector machines [3, 16].

Masters (2009) suggested collecting quadrographs for delay and trigraphs for duration. In this case, the information is stored in a special matrix. This avoids restoring the chronological history of keystrokes. It improves data privacy.

Studies also show authentication accuracy for relaxed and weary users [17, 18]. They showed that 70% of users reduce typing speed while in a bad emotional state and 83% of users increase typing speed.

Various types of systems are known in which the registration procedure may be different. Systems may specialize in entering the same fixed key sequence multiple times. The amount of input data can vary from five [19] to more than one hundred [19]. The obtained data are compared with the biometric model formed in the system.

For biometric identification by the dynamics of keystrokes, a lot of attention is paid to the keyboard used and sensors of biometric characteristics. In [20], such devices were compared. On different keyboards, the dynamics of keystrokes are different.

As the main sensor in such a system is a temporal timer. Such a timer for each programming language is different [21]. An external timer is also used [6, 11, 23]. Such studies used short passphrases. The use of long textual key sequences was also investigated [8]. Studies were carried out using a group of keys to increase the accuracy of identification [22, 23] on freely formed symbol sequences.

Some systems use a keystroke to build a biometric model. For this, pressure sensors are used [9, 24–26]. A sudden motion sensor [27] and sound signals [28] are also used. Many additional sensors together improve the technical characteristics of the identification system.

A lot of work is devoted to using the keyboard on a mobile phone [13, 29, 30], as well as the use of a touch keyboard.

Studies were conducted that showed the ability to determine the location of the keys on the keyboard [31].

Important research is the possibility of realizing a dialogue between the system and the user [32]. The system prompts the user to enter arbitrary text, which is evaluated by her.

Almost every system implements the selection of characteristic features. It is important that their processing speed be high [33]. Characteristics for hard-pressed keys were taken into account [34]. Basically, the complexity is characterized by the location of the keys (the distance between the keys), which are adjacent in the dynamics of their pressing, as well as the use of simultaneous pressing of several keys to form a key symbol (for example, using a key "Shift").

A lot of work is devoted to the extraction of features, which are determined by analyzing the primary characteristic features obtained from biometric keyboard sensors. These include minimum, maximum, and average values, as well as text slope, entropy, and spectral information [35, 36]. The normalization function [37] is also used to evaluate the normal distribution of values. To date, many methods have been developed aimed at the analysis of selected features [38, 39].

Existing decision making methods for expected user compliance are based on statistical methods and many different intellectual analysis methods [16, 19, 40, 41]. These methods use the calculation of the mean vector and standard deviation, the K method of nearest neighbors, cluster methods, neural network approaches, etc.

To increase the performance of the biometric identification process, a method is used based on the formation of a request for the introduction of a key sequence several times [15, 21, 41]. This procedure increases the probability of correct identification. An update of the biometric identification model during the functioning of the system is also used [14, 41]. However, it is possible to add impostor data. A good way to increase productivity is to combine several models of biometric identification [42]. In [43], three different methods are used for the dynamics of keystrokes.

To date, there is no reliable methodology for assessing the benefits of a method. Typically, performance, accuracy, and safety ratings exist.

To compare the system dynamics of keystrokes, several existing databases are used. At the same time, a good database must exactly meet various requirements [44]. The database should take into account age, culture, and other features, which is very difficult to take into account. Also, for a comparative assessment of the biometric model, indicators such as FAR, FRR, FMR, and FNMR. Each of these indicators has its own allowable numerical values. A receiver operating curve index (ROC) is also used. This curve is also known as the error curve.

A lot of attention was paid to working with different time intervals between the various pressed keys in the sequence (digraphs, trigraphs, etc.), and attempts were made to study for various contexts of a freely formed key sequence [45, 46]. In [47], studies with fixed and free key sequences were combined. However, the use of a free key sequence did not give a high percentage of accurate identification. From 4 to 30 users were involved in the study. In [48], test results from 40 to 165 users were presented. The results were very good. A larger number of test takers were also used.

Of great importance are methods that use preliminary training. Methods based on reference vector machines and neural networks show good results. However, it is difficult to implement and constantly maintain. Methods based on decision trees are used and developed [49, 50], fuzzy logic methods [51], genetic algorithms and other methods of intelligent data processing [19, 52, 53]. However, methods that use learning do not produce the desired result when a free key sequence is used.

## 4 Biometric Identification Model Based on Keystroke Dynamics

Keystroke dynamic identification model is a model that describes the process of extracting and generating biometric features and assigning them to a given class of signals generated at the keyboard output as a result of user work. In accordance with the selected model, a structure of characteristic features is formed, as well as methods for their extraction.

Today, there are a large number of models on the basis of which user identification based on keyboard dynamics is implemented. Almost all existing models are based on the use of characteristic features represented by time intervals (time of keystroke, time between keystrokes, etc.). An invariable component is the key sequence used. It can be fixed or randomly formed. Using a fixed key sequence (FKS) simplifies the database and identification model. Arbitrary key sequence (AKS) is arbitrarily user-generated. The identification based on AKS complicates the database, and the identification model is also complicated. Many identification methods are difficult to apply to this model.

To build a biometric model, information is collected. A well-known user performs the formation of the necessary key sequences on the keyboard. The sensors used respond to the keystroke dynamics and form a sequence of numerical values that are stored in memory.

When information is collected from the obtained numerical sequences, the quantitative characteristics of the necessary characteristic features are extracted and calculated in accordance with the selected model. From the received data the reference database of each user are formed. This stage is a system learning stage.

The next step is the classification step. At this stage, the resulting sequence is compared at the input with the reference sequences. For the implementation of this stage, different classification methods can be used.

The last stage is the stage at which the analysis of the received data at the current moment and previous moments of identification is carried out.

In addition, a step for updating user templates is possible. Updating of templates can occur either by adding additional data or by modifying an existing standard.

The proposed model should take into account the following requirements.

1. Possibility of software and hardware implementation.
2. Minimum number of uniform characteristic features.
3. Ability to engage a large number of users represented by one model.
4. High speed identification.
5. High accuracy of identification.

The first requirement indicates that the model used should be acceptable for the implementation of its software and hardware. A software-implemented model can form and use a large database, as well as use a large number of identification methods. The hardware implementation is based on the use of a circuitry structure that is adequate to the proposed model of biometric identification. The hardware-implemented model uses a small number of characteristic features and a small number of users. The hardware model is best used to verify a single user. Although a large database can be connected externally.

The first requirement forces us to develop a model that uses a small number of characteristic features and be simple. The number of bits of the device determines the number of identifiable users.

Homogeneous characteristic features have one physical nature. For keystroke dynamics, one parameter is used—time. Therefore, methods and means of measuring time intervals are used to extraction characteristic features. As a rule, these are counters of various modifications. A number represents time. Numbers in the construction of the model can be used in different numbers. The remaining requirements are inherent in all methods of biometric identification.

In accordance with the specified requirements, the identification model is described as follows.

1. Users who participate in the identification process form a template database. To do this, they form fixed and arbitrary key sequences at different times of the day.
2. The analysis of the generated key sequences is carried out and the necessary characteristic features are extracted.
3. In the identification process, the user enters a passphrase. The passphrase can be fixed text or free text. If a fixed passphrase is entered and it closely approaches one of the reference ones without exceeding a predetermined threshold value, then the user is identified.

If the fixed key phrase differs from the template ones and exceeds the specified threshold in quantitative values, then the analysis of differences from the nearest template sequences. It is assumed that due to various circumstances (emotional state, illness, etc.), the rhythm of the user's work behind the keyboard has broken. In this case, additional characteristic features are used, formed for each user individually. The system selects these characteristics for the closest user. In accordance with the analysis of deviations from all additional characteristic features are selected those that are control for this deviation. The system suggests introducing additional fixed key sequences that are stable for a given user. If the entered additional key sequences do not exceed the specified threshold, then the user is identified correctly. This sequence of events is carried out for all nearby users. The closest user is the user whose key template sequence exceeds the threshold for correct identification, but does not exceed the proximity threshold.

4. One of the possible modes of the model is the data update mode. This possibility exists because the user can enter an arbitrary key sequence. However, data is updated only if the system is confident (100% identification) that it is not an impostor, but the right user. In this case, the quantitative values of previously generated characteristic features are adjusted or new characteristic features are added.

The most important stage of intelligent data processing is the formation of characteristic features and their quantitative characteristics for each user in accordance with a given model.

To form the database of each user, a fixed key sequence (FKS) and an AKS are used. FKS is selected so that the following key combinations are formed.

- All keyboard keys pressed;
- All pairs of key combinations are pressed as two adjacent keys;
- All key combinations are pressed, three keys, etc.;
- The same key combinations should occur at the beginning, in the middle and at the end of the key sequence.

Moreover, to implement the formation of a fixed key sequence in which the necessary combinations are realized, a key sequence of large length is needed. Also, a fixed key sequence should be well remembered in order to recreate the user's keystroke dynamic.

To fix this problem, several FKSs with different combinations of adjacent keys can be used. FKS are entered several times, as a result of which the interval of spread of numerical values for each key sequence is calculated and the obtained acceptable interval is stored in the user database. Numerical values for the most stable adjacent key combinations in all key sequences are also determined and allowable intervals are calculated.

Using an AKS allows you to determine the numerical values for the key combinations present in the generated key sequence. The obtained numerical values of the key combinations are compared with the numerical values of the key combinations are compared with the numerical values of the combinations obtained from the generated fixed key sequences. The most stable numerical values of key combinations are selected and entered into the template database for each user.

For example, after generating all key sequences, numerical values of time intervals for all keys and their combinations were determined. Among all the combinations, combinations of keys are defined that form equal numerical values in cases of their different locations.

For example,

$$t^1_{AB} = t^2_{AB} = \ldots = t^n_{AB}$$

$$t^1_{NC} = t^2_{NC} = \ldots = t^{n.}_{NC}$$

These combinations are within a specific, given numeric interval ($t_{AB,min}$; $t_{AB,max}$) and ($t_{NC,min}$; $t_{NC,max}$). This interval can only be precisely selected experimentally. For each user, the confidence interval may be different.

Thus, as a result, a user database is formed, which contains the following arrays of numbers.

1. Sequences of numbers representing fixed key sequences.

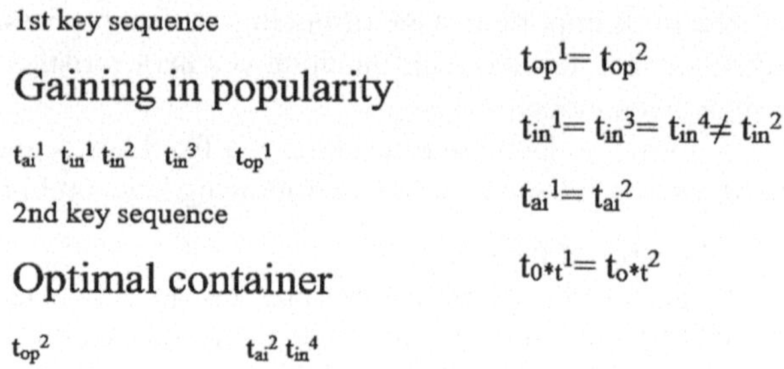


**Fig. 1** An example of the choice of numerical values for the same pairs of combinations of adjacent keys

2. Sequences of numbers that display stable values of time intervals between keystrokes of the same keys. Moreover, the location of these combinations can be placed anywhere in the key sequence (Fig. 1).

From this example it can be seen that the $t_{op}$ and $t_{ai}$ values are selected, which determine the time interval between keystrokes of the O and T keys, which are separated by various single characters, indicated by crosses in the indices between the O and P characters. The time between keystrokes of these keys falls into the confidence interval and therefore also falls into the generated database of this user.

In fact, characteristic features are extracted in accordance with the intellectual analysis of the generated key sequences and for each user they can be different. After the database is formed, the system is ready for identification.

In the identification mode, the user enters one of the FKSs proposed to him or the user forms the AKS of unlimited length. The longer the generated key sequence, the more accurately the user can be identified.

After the user has generated a key sequence at the system input, this sequence is analyzed. The system can operate in the FKS and AKS mode. If the sequence is fixed, then the "nearest" template key sequence is searched in the system database.

The template key sequence is "closest" to the entered key sequence, the number of its characteristic feature (CF) coinciding is greater than that of the rest of the template database key sequences. If the key sequence is 100% identical to one of the template ones, then the user is 100% identified. If there is a mismatch in the FKS with the nearest template key sequence (TKS), then the system uses the database of the closest user, which stores additional CFs. In this case, those additional CFs are selected that are not contained in the input FKS or are among the non-coincident numerical values. As a result of this analysis, the system prompts the user at the input to enter an additional key sequence (AKS) that contains additional characteristic features (ACF).

When the user enters an additional FKS, the system analyzes the generated ACFs and compares them with the template ones for the selected user from the generated database. If the quantitative values of ACF coincide, then the system gives a positive identification result; otherwise, a negative result is formed. The diagram of the user identification process in the FKS mode on Fig. 2 is shown.

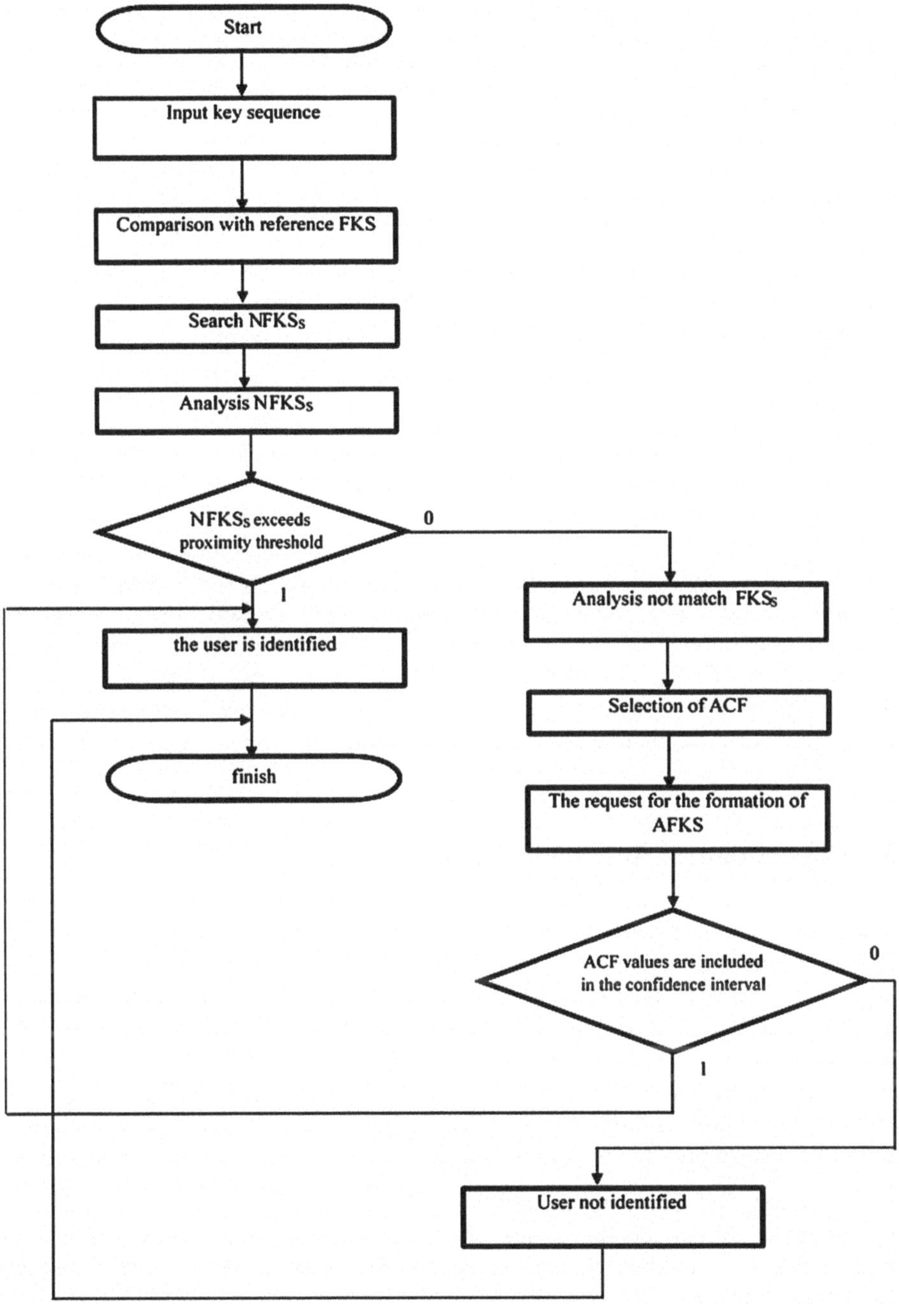


**Fig. 2** The flowchart of the user identification process in the FKS mode

The analysis block of the "nearest" $FKS_E$ requires the development of special algorithms for intelligent data processing for the proposed model of biometric identification.

Analysis blocks also require intelligent data processing if the parts of the $FKS_E$ P and the selection of ACF do not match.

As a result of the analysis of the "closest" $FKS_E$, the numbers of the input sequence are compared with the numbers of all $FKS_E$. In this case, the numbers located in a certain place of the FKS are compared with each other. For example, the first number of FKS is compared with the first number of FKS, the second number of FKS is compared with the second number of $FKS_E$, etc.

FKS is considered identified if the following condition is met

$$N \geq T, \quad (1)$$

where $N$—the number of matched numbers in FKS and $FKS_E$;

$T$—a number that determines the proximity of the $FKS_E$ to the identified FKS.

If condition (1) is not satisfied ($FKS_E$ is the "closest"), then the analysis of non-matching members of the $FKS_E$ is carried out. The key combinations of the mismatched part are determined, and from the database of the "closest" $FKS_E$ ($CFKS_E$), an additional $FKS_E$ ($AFKS_E$) is selected, which contains as many mismatched combinations as possible. Also can be select multiple $AFKS_E$ from the selected database.

The system offers the user FKS selected from the "nearest" database. However, the user may not be identified, and his database is present in the system. In this situation, the database update or addition mode is used.

## 5 Experimental Research Results

Thirty people were invited to participate in the experiment, who for a long time entered text into the computer system using the keyboard; each participant in the experiment introduced the same FKS in the mother tongue ten times into the computer system. Each test person entered FKS in the language that he thinks and that he is fluent.

The entered text was recorded by a special program, and the program recorded all time characteristics during the introduction of text into the system. Thus, at the end of the FKS input, the program formed:

- the sequence of characters entered;
- a sequence of numbers that correspond to the length of time the keystrokes ($T_i$);
- a sequence of numbers that correspond to the durations between two pressed adjacent keys ($T_{i,\,i+1}$);
- a sequence of numbers that correspond to the time of the trigraphs ($T_{i,i+2}$), etc.

As a result of the program, an array of triangular-shaped numbers is formed (Fig. 3).

ffjflkkflsmcksacnajsknv

| № n/n | SYMBOL | 0,15 | t_+1 | t_+2 | t_+3 | t_+4 | t_+5 | t_+6 | t_+7 | t_+8 | t_+9 | t_+10 | t_+11 | t_+12 | t_+13 | t_+14 | t_+15 | t_+16 | t_+17 | t_+18 | t_+19 | t_+20 | t_+21 | t_+22 |
|---|---|---|---|---|---|---|---|---|---|---|---|---|---|---|---|---|---|---|---|---|---|---|---|---|
| 1 | f | 0,127 | 0 | 0 | 0 | 0 | 0 | 0 | 0 | 0 | 0 | 0 | 0 | 0 | 0 | 0 | 0 | 0 | 0 | 0 | 0 | 0 | 0 | 0 |
| 2 | f | 0,088 | 0,302 | 0 | 0 | 0 | 0 | 0 | 0 | 0 | 0 | 0 | 0 | 0 | 0 | 0 | 0 | 0 | 0 | 0 | 0 | 0 | 0 | 0 |
| 3 | j | 09:41:1 | 0,142 | 0,444 | 0 | 0 | 0 | 0 | 0 | 0 | 0 | 0 | 0 | 0 | 0 | 0 | 0 | 0 | 0 | 0 | 0 | 0 | 0 | 0 |
| 4 | f | 0,094 | 0,048 | 0,19 | 0,492 | 0 | 0 | 0 | 0 | 0 | 0 | 0 | 0 | 0 | 0 | 0 | 0 | 0 | 0 | 0 | 0 | 0 | 0 | 0 |
| 5 | l | 09:41:1 | 0,15 | 0,198 | 0,34 | 0,642 | 0 | 0 | 0 | 0 | 0 | 0 | 0 | 0 | 0 | 0 | 0 | 0 | 0 | 0 | 0 | 0 | 0 | 0 |
| 6 | k | 0,139 | 0,001 | 0,151 | 0,199 | 0,341 | 0,643 | 0 | 0 | 0 | 0 | 0 | 0 | 0 | 0 | 0 | 0 | 0 | 0 | 0 | 0 | 0 | 0 | 0 |
| 7 | k | 09:41:1 | 0,194 | 0,195 | 0,345 | 0,393 | 0,535 | 0,837 | 0 | 0 | 0 | 0 | 0 | 0 | 0 | 0 | 0 | 0 | 0 | 0 | 0 | 0 | 0 | 0 |
| 8 | f | 09:41:1 | 0,023 | 0,217 | 0,218 | 0,368 | 0,416 | 0,558 | 0,86 | 0 | 0 | 0 | 0 | 0 | 0 | 0 | 0 | 0 | 0 | 0 | 0 | 0 | 0 | 0 |
| 9 | l | 0,002 | 0,2 | 0,223 | 0,417 | 0,418 | 0,568 | 0,616 | 0,758 | 1,06 | 0 | 0 | 0 | 0 | 0 | 0 | 0 | 0 | 0 | 0 | 0 | 0 | 0 | 0 |
| 10 | s | 0,217 | 0,019 | 0,219 | 0,242 | 0,436 | 0,437 | 0,587 | 0,635 | 0,777 | 1,079 | 0 | 0 | 0 | 0 | 0 | 0 | 0 | 0 | 0 | 0 | 0 | 0 | 0 |
| 11 | m | 09:41:1 | 0,218 | 0,237 | 0,437 | 0,46 | 0,654 | 0,655 | 0,805 | 0,853 | 0,995 | 1,297 | 0 | 0 | 0 | 0 | 0 | 0 | 0 | 0 | 0 | 0 | 0 | 0 |
| 12 | c | 0,083 | 0,002 | 0,22 | 0,239 | 0,439 | 0,462 | 0,656 | 0,657 | 0,807 | 0,855 | 0,997 | 1,299 | 0 | 0 | 0 | 0 | 0 | 0 | 0 | 0 | 0 | 0 | 0 |
| 13 | k | 0,002 | 0,191 | 0,193 | 0,411 | 0,43 | 0,63 | 0,653 | 0,847 | 0,848 | 0,998 | 1,046 | 1,188 | 1,49 | 0 | 0 | 0 | 0 | 0 | 0 | 0 | 0 | 0 | 0 |
| 14 | s | 09:41:1 | 0,002 | 0,193 | 0,195 | 0,413 | 0,432 | 0,632 | 0,655 | 0,849 | 0,85 | 1 | 1,048 | 1,19 | 1,492 | 0 | 0 | 0 | 0 | 0 | 0 | 0 | 0 | 0 |
| 15 | a | 0,1 | 0,002 | 0,004 | 0,195 | 0,197 | 0,415 | 0,434 | 0,634 | 0,657 | 0,851 | 0,852 | 1,002 | 1,05 | 1,192 | 1,494 | 0 | 0 | 0 | 0 | 0 | 0 | 0 | 0 |
| 16 | c | 09:41:1 | 0,128 | 0,13 | 0,132 | 0,323 | 0,325 | 0,543 | 0,562 | 0,762 | 0,785 | 0,979 | 0,98 | 1,13 | 1,178 | 1,32 | 1,622 | 0 | 0 | 0 | 0 | 0 | 0 | 0 |
| 17 | n | 0,17 | 0,019 | 0,147 | 0,149 | 0,151 | 0,342 | 0,344 | 0,562 | 0,581 | 0,781 | 0,804 | 0,998 | 0,999 | 1,149 | 1,197 | 1,339 | 1,641 | 0 | 0 | 0 | 0 | 0 | 0 |
| 18 | a | 09:41:1 | 0,17 | 0,189 | 0,317 | 0,319 | 0,321 | 0,512 | 0,514 | 0,732 | 0,751 | 0,951 | 0,974 | 1,168 | 1,169 | 1,319 | 1,367 | 1,509 | 1,811 | 0 | 0 | 0 | 0 | 0 |
| 19 | j | 09:41:1 | 0,026 | 0,196 | 0,215 | 0,343 | 0,345 | 0,347 | 0,538 | 0,54 | 0,758 | 0,777 | 0,977 | 1 | 1,194 | 1,195 | 1,345 | 1,393 | 1,535 | 1,837 | 0 | 0 | 0 | 0 |
| 20 | s | 09:41:1 | 0,002 | 0,028 | 0,198 | 0,217 | 0,345 | 0,347 | 0,349 | 0,54 | 0,542 | 0,76 | 0,779 | 0,979 | 1,002 | 1,196 | 1,197 | 1,347 | 1,395 | 1,537 | 1,839 | 0 | 0 | 0 |
| 21 | k | 0,048 | 0,03 | 0,032 | 0,058 | 0,228 | 0,247 | 0,375 | 0,377 | 0,379 | 0,57 | 0,572 | 0,79 | 0,809 | 1,009 | 1,032 | 1,226 | 1,227 | 1,377 | 1,425 | 1,567 | 1,869 | 0 | 0 |
| 22 | n | 09:41:1 | 0,099 | 0,129 | 0,131 | 0,157 | 0,327 | 0,346 | 0,474 | 0,476 | 0,478 | 0,669 | 0,671 | 0,889 | 0,908 | 1,108 | 1,131 | 1,325 | 1,326 | 1,476 | 1,524 | 1,666 | 1,968 | 0 |
| 23 | v | 0,047 | 0,002 | 0,101 | 0,131 | 0,133 | 0,159 | 0,329 | 0,348 | 0,476 | 0,478 | 0,48 | 0,671 | 0,673 | 0,891 | 0,91 | 1,11 | 1,133 | 1,327 | 1,328 | 1,478 | 1,526 | 1,668 | 1,97 |

**Fig. 3** An example of a formed array after entering one FKS

In the last column of the generated array there is one number that determines the time of FKS entry into the system.

All generated FKS of one user were compared among themselves. A confidence interval was selected that specified the coincidence of numbers. For evaluation, the following condition was used

$$\left(T_i^1 - T_i^2\right) = \begin{cases} 1, if \left|T_i^1 - T_i^2\right| < A \\ 0, in\ othe\ case \end{cases}, \qquad (2)$$

where A—the boundary of the permissible deviation of the $T_i$ values of two FKS. The same condition (2) is used for D-graphs, trigraphs, etc.

As a result of comparing the quantitative values of each FKP of the same user, triangular arrays were formed in which the units indicated the same values within the accepted confidence interval. Zeros indicate mismatched values within the same confidence interval. Also, the percentage of matches was determined for each array of homogeneous quantities (for example, for D-graphs). An example of the generated arrays for FKS generated by one user on Fig. 4 is presented.

Some close values for D-graphs are possible. However, for trigraphs, etc. significant differences are observed. Situations when trigraphs have close meanings and D-graphs differ are quite rare.

As can be seen from the presented arrays of matches of the same user, they have a high percentage of matches of homogeneous quantities. Moreover, the distribution of units in each homogeneous coincidence array for different FKSs has a different structure. However, the percentage of matches for different FKSs of one user does not have sharp differences. Significant differences are observed for the FKS of different users (Fig. 5).

Studies were conducted for different confidence intervals of one user and different users. Figure 6 shows the dependencies of coincidence arrays for one user and for different users for different confidence intervals.

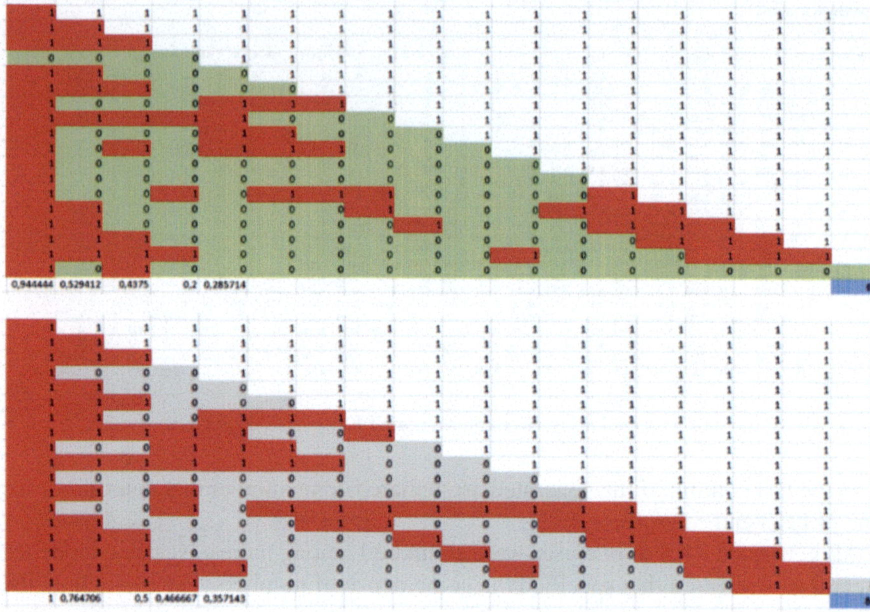


**Fig. 4** Examples of coincidence arrays for several FKS of the same user

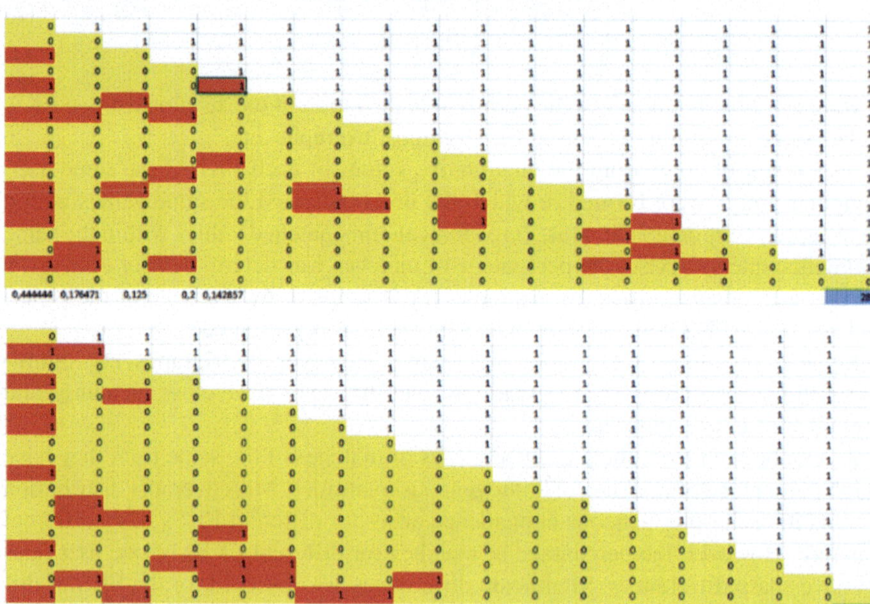


**Fig. 5** Examples of arrays of FKS matches generated by different users

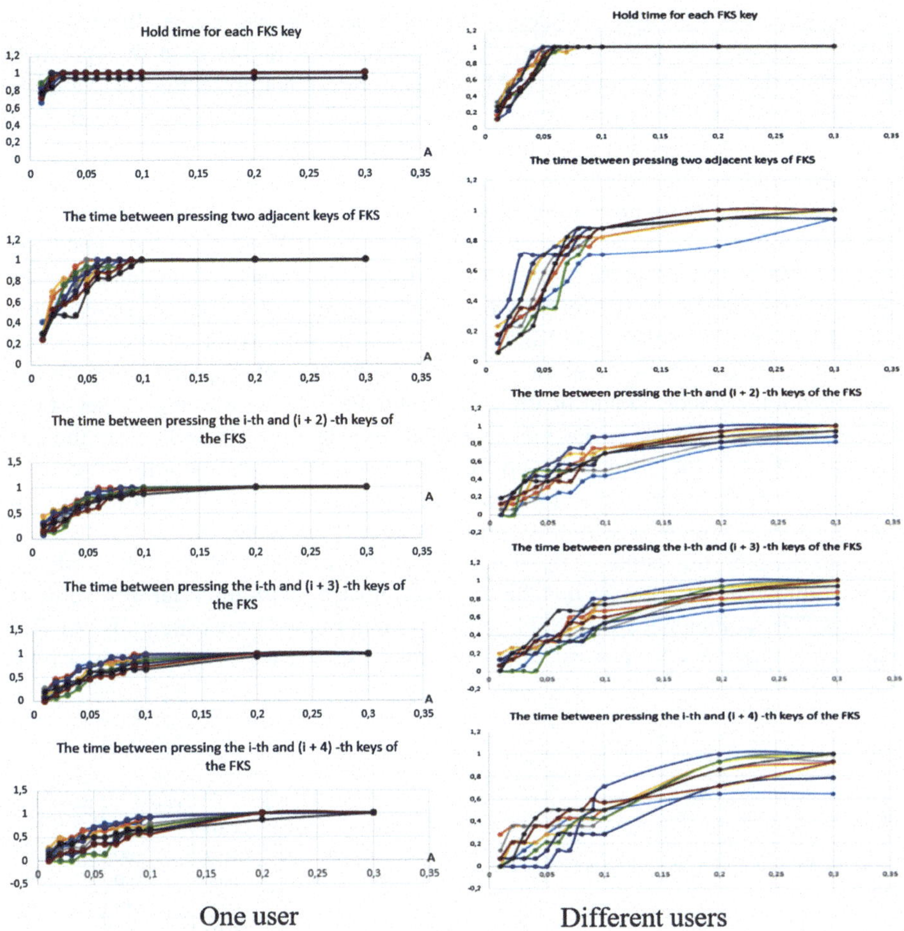


**Fig. 6** Dependencies of coincidence arrays for one user and for different users for different confidence intervals

Figure 6 shows the first five keyboard keys and time characteristics for them. However, FKS has more characters. The color represents the corresponding time intervals between keystrokes and for the keys themselves. As can be seen from the figure for different characteristics, complete coincidence occurs for different confidence intervals. However, you can choose the necessary confidence optimal interval.

The experiment involved the same user whose characteristics were compared with another user. Studies showed a high percentage of matches (above 95%) for a single user at intervals of 0.1. Studies for different users showed a low percentage of matches on almost all the studied confidence interval values.

If you examine each characteristic feature, represented by a separate column in the match table, you can select the appropriate percentage of matches that will distinguish the user from the impostor. For this, five characteristic features were investigated. The required confidence interval was also selected. A difference of 10% was chosen for identification. If the difference is higher than the specified value, then the user is not identified. Moreover, for each feature, a separate value of the coincidence threshold was chosen. For example, for the third characteristic feature, the threshold value is higher than for the fourth.

Based on certain threshold values adopted for each characteristic, the values of FRR (false denial of identification) and FAR (false identification of someone else) were determined. The better the system, the lower the FRR value for the same FAR values. Characteristic features were used starting from the second and ending with the fifth obtained during the formation of the histogram. The first five signs determining the time of pressing and holding the key, as well as the time of pressing between other keys, were taken into account. In the case of disputed values at the boundary of the selected identification threshold, the time of the complete set of the key sequence and the last two values that determine the time elapsed before the last two characters of the sequence were pressed were also used. Graphs were constructed for FRR and FAR for different confidence intervals and for different percentages for coincidences with the template values (Figs. 7 and 8).

From the presented studies, the use of the confidence interval A = 0.09 with the optimal threshold for percent matches when comparing with the reference value is obvious. The coincidence threshold value was determined experimentally. At this confidence interval, both control values are zero. In this case, A = 0.08 can also be

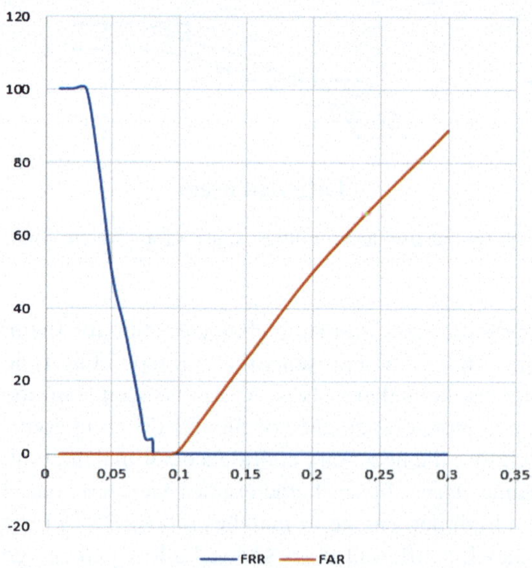


| A | FRR | FAR |
|---|---|---|
| 0,01 | 100 | 0 |
| 0,02 | 100 | 0 |
| 0,03 | 100 | 0 |
| 0,04 | 78 | 0 |
| 0,05 | 49 | 0 |
| 0,06 | 33 | 0 |
| 0,07 | 14 | 0 |
| 0,075 | 4 | 0 |
| 0,08 | 4 | 0 |
| 0,0805 | 0 | 0 |
| 0,081 | 0 | 0 |
| 0,082 | 0 | 0 |
| 0,085 | 0 | 0 |
| 0,09 | 0 | 0 |
| 0,1 | 0 | 1 |
| 0,2 | 0 | 50 |
| 0,3 | 0 | 89 |

**Fig. 7** FRR and FAR plots for confidence intervals from 0.01 to 0.3 and with the highest threshold of coincidence with reference values

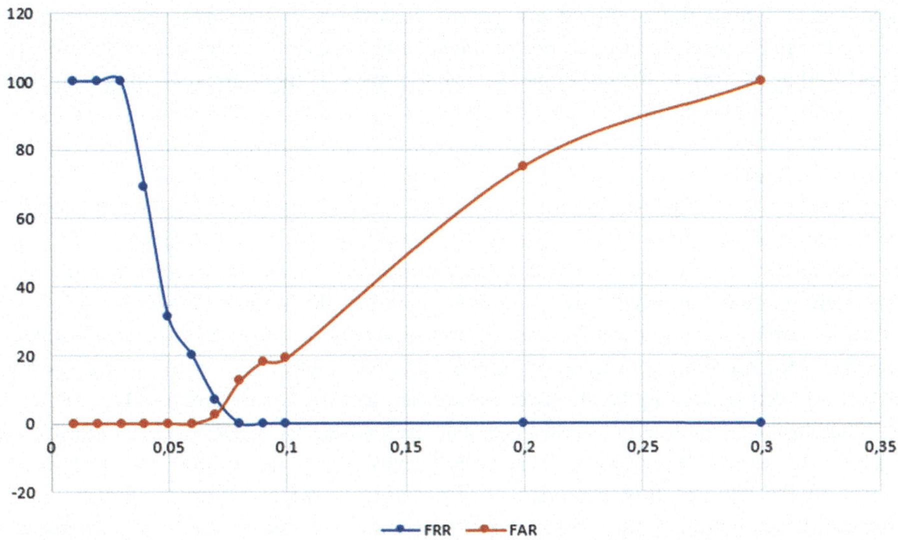


**Fig. 8** FRR and FAR plots for confidence intervals from 0.01 to 0.3 and with an average threshold of coincidence with reference values

used, since at this value FAR = 0% and FRR = 4%, which is an acceptable value. The experiment involved users who for a long time worked with a computer keyboard.

Using the obtained experimental data, the values of the number of matches for different time values were determined after comparison with the template values. Ten template values were used for each user for each key sequence. In this case, the reference values were used of different lengths. If controversial results were obtained that do not allow for confident identification, the system offered additional sequences that the user entered.

If one of the parameters is slightly lower than the specified threshold, and the remaining parameters following it, significantly exceed the threshold value, then a decision is made on the person identification. This approach in the experiment did not give false identification results and did not increase the value of FAR.

The described approach gives more accurate results when reference arrays are used as much as possible, which requires the use of additional resources. Also a lot depends on the mood or physical state of the user, which affect the uniformity of text input.

More stable results are given by the method, which is as follows.

Form the base of reference arrays. There should be at least ten template arrays. When entering the key text, the resulting array is compared with all formed arrays and the total percentage of comparisons for all arrays are got. The threshold value for the received percentage of matches is set. The established confidence interval is selected at which the obtained arrays are compared. If the obtained percentage exceeds a predetermined threshold for the selected reference values, then the user is identified according to the generated template arrays. If there are several options for exceeding the threshold, then an additional two parameters of each array are used, which determine the maximum times for typing a key phrase and are formed at the end of each array.

For this method, threshold coincidence values were also established and a confidence interval was determined. Also was studied key sequences of different lengths of characters, and also used text sequences and sequences consisting of numbers. The ratio of FRR and FAR without a selected threshold in Fig. 9 is shown.

Both curves intersect for 25% of errors, which allows you to choose a reliable threshold for reliable identification. At the selected threshold, 50% of the graphs for FRR and FAR according to the second method have the form shown in Fig. 10.

The choice of such a threshold gives a section where the FRR and FAR curves in a particular section have a value of 0, which gives confidence in the correct identification of the user. In this approach, it is possible to select different identification thresholds, coincidence thresholds, confidence interval and threshold in percentage limit. For longer FKSs, the dependency graphs for FRR and FAR differ in shape. The confidence interval narrows (Fig. 11).

The analysis shows that for each selected FKS, the confidence interval is shifted. Therefore, during identification, it is necessary to take into account and set the confidence interval and threshold values, taking into account the FKS used.

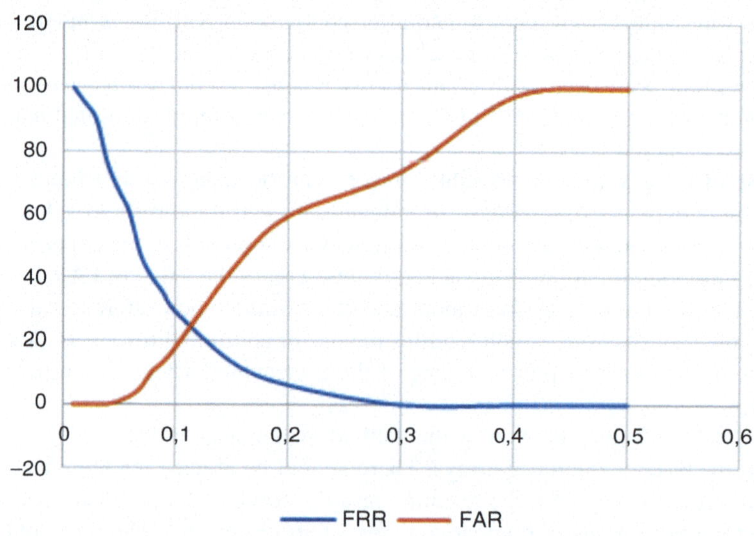


**Fig. 9** FRR and FAR plots according to the second method with percentage determination of matches for a sequence of 12 ordered numbers

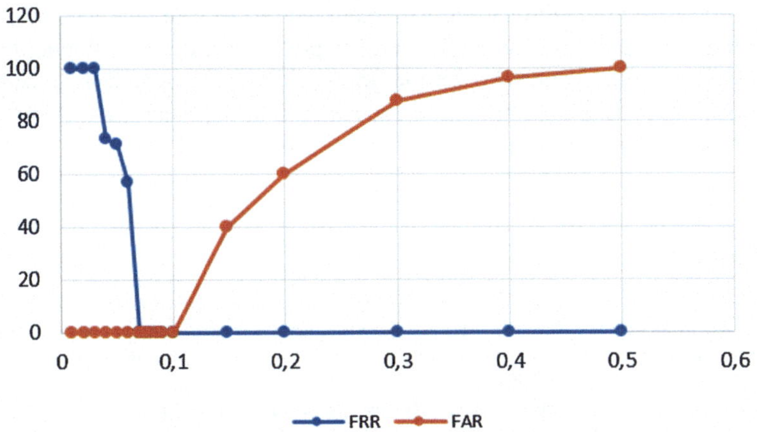


**Fig. 10** FRR and FAR plots for the second method with a 50% percent identification threshold for a sequence of 12 ordered numbers

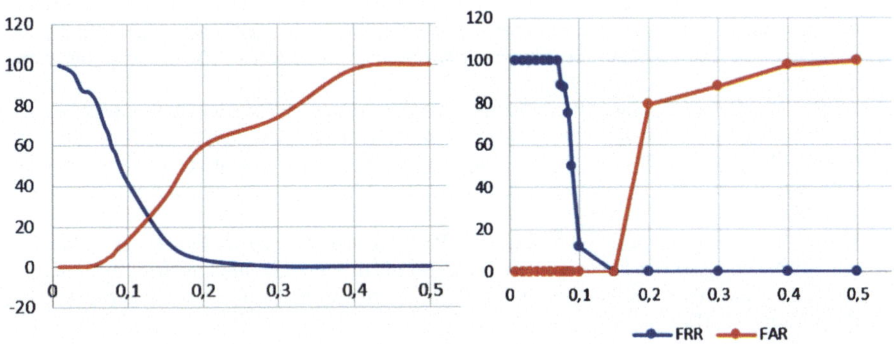


**Fig. 11** FRR and FAR plots according to the second method with a 50% percent identification threshold for a sequence of 18 ordered text characters

## 6 Conclusion

The paper presents a technique for identifying and authenticating a person using keystroke dynamic. The presented forms of template arrays for each fixed key sequence made it possible to automatically analyze the temporal characteristics of the keystroke handwriting of a person. The use of five threshold values of the temporal characteristics of handwriting, as well as two additional characteristics, made it possible to identify a person with high accuracy. Based on the experiments, the necessary threshold values were selected at which the FRR and FAR values tend to zero, which does not lead to a possible false identification. The stability of keystroke handwriting for various times of the day was also evaluated. The proposed technique allows us to constantly increase the template base for each user with each session of successful identification, which allowed us to take into account its continuous improvement. Variation in the percentage of coincidence with template values allowed us to determine the optimal threshold values, as well as to select the necessary numerical characteristics for each user. The technique used allows you to identify a user who does not often use the keyboard.

## References


1. Hobi, M., Majeed, M., & Sarab, H. (2013). Biometrics for user authentication using artificial neural networks. In *Keystroke dynamics for user authentication*. Buffalo, NY: LAP LAMBERT Academic Publishing.
2. Dharmendra, S., Bhawnesh, J., Himanshu, N., & Amit, K. (2017). Presskey- a keystrokes dynamics based authentication system. *International Journal of Advanced Research in Computer Science, 8*(5), 2665–2670.
3. Giot, R., El-Abed, M., Hemery, B., & Rosenberger, C. (2011). Unconstrained keystroke dynamics authentication with shared secret. *Computers & Security, 30*(6-7), 1–20.
4. Patil, R. A., & Renke, A. L. (2016). Keystroke dynamics for user authentication and identification by using typing rhythm. *International Journal of Computer Applications, 144*(9), 27–33.
5. Avramidis, L. (2014). *Keystroke dynamic authentication as a service*, MA thesis, Harokopio University.
6. Azzini, A., Marrara, S., Sassi, R., & Scotti, F. (2008). A fuzzy approach to multimodal biometric continuous authentication. *Fuzzy Optimization and Decision Making, 7*(3), 243–256.
7. de Melo, L. J., & Vale, H. M. C. (2017). Improvement of security systems by keystroke dynamics of passwords. *International Journal of Computer Science and Information Security, 15*(9), 156–159.
8. Harun, N., Woo, W. L., & Dlay, S. S. (2010). Performance of keystroke biometrics authentication system using artificial neural network (ANN) and distance classifier method. In *Computer and Communication Engineering (ICCCE), 2010 International Conference on* (pp. 1–6). Piscataway, NJ: IEEE.
9. Killourhy, K. S. (2012). *A scientific understanding of keystroke dynamics*. PhD thesis, Carnegie Mellon University.
10. Miluzzo, E., Varshavsky, A., Balakrishnan, S., & Choudhury, R. R. (2012). Tapprints: your finger taps have fingerprints. In *Proceedings of the 10th International Conference on Mobile Systems, Applications, and Services* (pp. 323–336). New York: ACM.
11. Marsters, J.-D. (2009). *Keystroke dynamics as a biometric*. PhD thesis, University of Southampton.
12. Monrose, F., & Rubin, A. (1997). Authentication via keystroke dynamics. In *Proceedings of the 4th ACM Conference on Computer and Communications Security* (pp. 48–56). New York: ACM Press.
13. Hocquet, S., Ramel, J.-Y., & Cardot, H. (2006). Estimation of user specific parameters in one-classproblems. In *ICPR'06: Proceedings of the18th International Conference on Pattern Recognition* (pp. 449–452). Washington, DC: IEEE Computer Society.
14. Revett, K. (2009). A bioinformatics based approach to user authentication via keystroke dynamics, international journal of control. *Automation and Systems, 7*(1), 7–15.
15. Bartmann, D., Bakdi, I., & Achatz, M. (2007). On the design of an authentication system based on keystroke dynamics using a predefined input text. *International Journal of Information Security and Privacy (IJISP), 1*(2), 149. Techniques and Applications for Advanced Information Privacy and Security: Emerging Organizational, Ethical, and Human Issues.

16. Giot, R., El-Abed, M., & Rosenberger, C. (2009). Keystroke dynamics with low constraints SVM based pass phrase enrollment. In *IEEE International Conference on Biometrics: Theory, Applications and Systems (BTAS 2009)* (pp. 1–6). Washington, DC: IEEE Computer Society.
17. Khanna, P., & Sasikumar, M. (2010). Recognising emotions from keyboard stroke pattern. *International Journal of Computer Applications, 11*(9), 24–28.
18. Giot, R., El-Abed, M., Hemery, B., & Rosenberger, C. (2011). Unconstrained keystroke dynamics authentication with shared secret. *Computers & Security, 30*(6), 427–445.
19. Obaidat, M., & Sadoun, B. (1997). Verification of computer users using keystroke dynamics. In *Systems, Man and Cybernetics, Part B, IEEE Transactions on* (Vol. 27(2), pp. 261–269). New York: Institute of Electrical and Electronics Engineers.
20. Ross, A., & Jain, A. (2004). Biometric sensor interoperability: A case study in fingerprints. In *Proc. of International ECCV Workshop on Biometric Authentication (BioAW)* (pp. 134–145). Berlin: Springer.
21. Killourhy, K., & Maxion, R. (2008). The effect of clock resolution on keystroke dynamics. In *Proceedings of the11th international symposium on Recent Advances in Intrusion Detection* (pp. 331–350). Berlin: Springer.
22. Huang, X., Lund, G., & Sapeluk, A. (2012). Development of a typing behaviour recognition mechanism on android. In *Trust, Security and Privacy in Computing and Communications (TrustCom), 2012 IEEE 11th International Conference on* (pp. 1342–1347). Los Alamitos, CA: IEEE.
23. Li, L., Zhao, X., & Xue, G. (2013). *Unobservable re-authentication for smartphones*. New York: NDSS.
24. Grabham, N., & White, N. (2008). Use of a novel keypad biometric for enhanced user identity verification. In *Instrumentation and Measurement Technology Conference Proceedings, 2008. IMTC 2008* (pp. 12–16). Piscataway, NJ: IEEE.
25. van der Loo, M. P., & de Jonge, E. (2013). *An introduction to data cleaning with R. tech rep Statistics, Netherlands*.
26. Ho, J., & Kang, Dae-Ki. (2015). Sequence alignment with dynamic divisor generation for keystroke dynamics based user authentication. Journal of Sensors 2015: 14.
27. Lopatka, M., & Peetz, M. (2009). Vibration sensitive keystroke analysis. In *Proceedings of the 18th Annual Belgian-Dutch Conference on Machine Learning* (pp. 75–80). Tilburg: Tilburg Center for Creative Computing.
28. Nguyen, T., Le, T., & Le, B. (2010). Keystroke dynamics extraction by independent component analysis and bio-matrix for user authentication. In B.-T. Zhang & M. Orgun (Eds.), *PRICAI 2010: Trends in Artificial Intelligence, Vol. 6230 of Lecture Notes in Computer Science* (pp. 477–486). Berlin: Springer.
29. Campisi, P., Maiorana, E., Lo Bosco, M., & Neri, A. (2009). User authentication using keystroke dynamics for cellular phones, signal processing. *IET, 3*(4), 333–341.
30. Epp, C., Lippold, M., & Mandryk, R. L. (2011). Identifying emotional states using keystroke dynamics. In *Conference a Human Factors in Computing Systems*. New York: ACM Press.
31. Araújo, L. C., Sucupira, L. H., Lizarraga, M. G., Ling, L. L., & Yabu-Uti, J. B. T. (2005). User authentication through typing biometrics features. *IEEE Transactions on Signal Processing, 53*(2 Part 2), 851–855.
32. Umphress, D., & Williams, G. (1985). Identity verification through keyboard characteristics. *International Journal of Man-Machine Studies, 23*, 263–273.
33. Balagani, K. S., Phoha, V. V., Ray, A., & Phoha, S. (2011). On the discriminability of keystroke feature vectors used infixed text keystroke authentication, pattern. *Recognition Letters, 32*(7), 1070–1080.
34. de Ru, W. G., & Eloff, J. H. P. (1997). Enhanced password authentication through fuzzy logic. *IEEE Expert: Intelligent Systems and Their Applications, 12*, 38–45.
35. Modi, S. K., & Elliott, S. J. (2006). Kesytroke dynamics verification using spontaneously generated password. In *IEEE International Carnahan Conferences Security Technology*. Piscataway, NJ: IEEE.

36. Chang, W. (2006). *Keystroke biometric system using wavelets, ICB 2006* (pp. 647–653). Berlin: Springer.
37. Filho, J. R. M., & Freire, E. O. (2006). On the equalization of keystroke timing histograms. *Pattern Recognition Letters, 27*, 1440–1446.
38. Boechat, G., Ferreira, J., & Carvalho, E. (2006). Using the keystrokes dynamic for systems of personal security, proceedings of world academy of science. *Engineering and Technology, 18*, 200–205.
39. Azevedo, G., Cavalcanti, G., Carvalho Filho, E., & Recife-PE, B. (2007). *An approach to feature selection for keystroke dynamics systems based on PSO and feature weighting, Evolutionary Computation, 2007. CEC 2007*. Piscataway, NJ: IEEE.
40. Sang, Y., Shen, H., & Fan, P. (2004). Novel impostors detection in keystroke dynamics by support vector machine. In *Proc. of the 5th International Conference on Parallel and Distributed Computing, Applications and Technologies (PDCAT 2004)*. Singapore: Springer.
41. Hosseinzadeh, D., & Krishnan, S. (2008). Gaussian mixture modeling of keystroke patterns for biometric applications. In *Systems, Man, and Cybernetics, Part C: Applications and Reviews, IEEE Transactions on* (Vol. 38(6), pp. 816–826). Piscataway: NY: IEEE.
42. Ross, A., Nandakumar, K., & Jain, A. (2006). *Handbook of Multibiometrics*. London: Springer.
43. Hocquet, S., Ramel, J.-Y., & Cardot, H. (2006). Estimation of user specific parameters in one-class problems. In *ICPR'06: Proceedings of the18th International Conference on Pattern Recognition* (pp. 449–452). Washington, DC: IEEE Computer Society.
44. Cherifi, F., Hemery, B., Giot, R., Pasquet, M., & Rosenberger, C. (2009). Performance evaluation of behavioral biometric systems. In *Behavioral biometrics for human identification: Intelligent applications* (pp. 57–74). Hershey, PA: IGI Global.
45. Buchoux, A., & Clarke, N. L. (2008). Deployment of keystroke analysis on a smartphone. In *Australian Information Security Management Conference* (p. 48). Boston: Springer.
46. Ilonen, J. (2003). Keystroke dynamics. In *Advanced Topics in Information Processing: 03–04*. Hershey, PA: Idea Group Publication.
47. Hwang, S.-s., Lee, H.-j., & Cho, S. (2009). Improving authentication accuracy using artificial rhythms and cues for keystroke dynamics-based authentication. *Expert Systems with Applications, 36*(7), 10649–10656.
48. Chudá, D., & Durfina, M. (2009). Multifactor authentication based on keystroke dynamics. In *Proceedings of the International Conference on Computer Systems and Technologies and Workshop for PhD Students in Computing* (p. 89). New York: ACM.
49. Saifan, R., Salem, A., Zaidan, D., & Swidan, A. (2016). A survey of behavioral authentication using keystroke dynamics: Touch screens and mobile devices. *Journal of Social Sciences, 5*(1), 29–41.
50. Crawford, H. (2010). Keystroke dynamics: Characteristics and opportunities. In *Privacy, Security and Trust (PST), 2010 Eighth Annual International Conference on* (pp. 205–212). Piscataway: NY: IEEE.
51. Eltahir, W. E., Salami, M. J. E., Ismail, A. F., & Lai, W. K. (2004). Dynamic keystroke analysis using AR model. In *Industrial Technology, 2004. IEEE ICIT'04. 2004 IEEE International Conference on* (Vol. 3, pp. 1555–1560). Piscataway: NY: IEEE.
52. Pavaday, N., & Nugessur, S. (2010). Investigating & improving the reliability and repeatability of keystrokedynamics timers. *International Journal of Network Security & Its Applications (IJNSA), 2*(3), 70–85.
53. Moskovitch, R., et al. (2009). Identity theft, computers and behavioral biometrics. In *Intelligence and Security Informatics, 2009. ISI'09. IEEE International Conference on* (pp. 155–160). Piscataway: NY: IEEE.

# Analysis of the Dynamics of Handwriting for Biometric Personality Identification Based on Cellular Automata

**Stepan Bilan, Mykola Bilan, Andrii Bilan, and Sergii Bilan**

## 1 Introduction

Currently, much attention is paid to handwriting analysis. First of all, it is important to know when directly text information entering into a computer system without using a keyboard. Handwritten text analysis is performed to solve the problems of recognition and identification of the person who forms the handwritten text. The solution to the recognition problem is that the meaning of the information is determined. Recognition can be performed both by statistical images and by the dynamics of text input using a writing element.

To solve the recognition problem, it is necessary to determine and recognize which symbol or group of symbols is formed. General characteristics for each symbol are defined here. Symbols or groups of characters can have different shapes, but have the same meaning for a computing system or for other people. Moreover, such a task can be difficult and its solution often gives false results, since the handwriting of different people can differ significantly from each other. The problem of handwriting recognition is devoted to many works [1–8]. The base of standards contains quantitative values of characteristic features that describe the geometric shape of a symbol or group of symbols.

In handwriting identification tasks, it is necessary to determine the personality that formed the text. However, the text itself is not always recognized. Identification by static images of handwritten text is mainly used in forensic science, in banking

S. Bilan (✉)
State University of Infrastructure and Technology, Kyiv, Ukraine
e-mail: bilan_sm@gsuite.duit.edu.ua

M. Bilan · A. Bilan
The Municipal Educational Institution Mayakskaya Secondary School, Mayak, Moldava

S. Bilan
Onseo Company, Vinnytsia, Ukraine



S. Bilan et al. (eds.), *Biometric Identification Technologies Based on Modern Data Mining Methods*, https://doi.org/10.1007/978-3-030-48378-4_4

and other areas where it is necessary to establish identity by signature or image of a text passage.

Modern access systems mainly use the identification of a person by the dynamics of handwriting input. As a rule, this is implemented using special tablets and writing elements. The dynamics of handwriting text input can be analyzed in real time, when the main characteristics are directly recorded and analyzed during input. In this mode, at a certain stage of input, it may deny access to the user, since the obtained value at a certain time step indicates the presence of an impostor.

The system can also, after entering the text, analyze all the obtained quantitative values of the characteristic features and make a decision on identification.

## 2 The Main Provisions of Biometric Identification of a Person According to the Dynamics of Handwritten Text

As mentioned earlier, the dynamics of handwriting formation is much more difficult to copy than a static image of handwritten text. This is due to the fact that all characteristic features are formed during the time of handwriting creation. Each quantitative value is determined by its place in the formed sequence of characteristic features.

The following characteristics are used to describe the dynamics of handwriting creation.

1. The formation of the entire control text.
2. The time for the formation of one character in handwriting.
3. The time intervals during which the writing element does not touch the surface of the application of handwriting.
4. Time intervals during which handwriting formation was accelerated.
5. Time intervals during which handwriting formation slowed down.

These characteristics are determined by time. However, geometrical characteristics, which include:

1. Symbol shape.
2. The geometric shape of a handwritten character group.

Using both groups of characteristic features allows to identify a person with high accuracy. Each person forms a symbol or a group of handwritten symbols of different geometric shapes. Moreover, in automatic identification systems, the task is to automatically highlight each individual character or other groups of characters. The separation of the symbol should be carried out in real time. Examples of spelling the word "happy" by different people are presented in Fig. 1.

Figure 1 shows the images of the geometric shape of the group of characters in handwritten text and the time it was written, as well as the time for writing the individual characters of the created text. In this case, individual quantitative values of

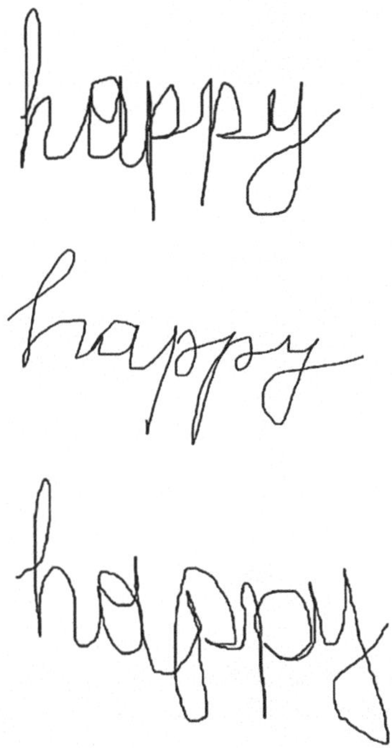

**Fig. 1** Examples of spelling the word "*happy*" by different people

the time of writing the symbol may coincide, but the forms in the details usually differ.

If you determine the increase in the number of added pixels that belong to the characters of the text, then the curve of dependence of the number of pixels of the text will be presented in Fig. 2.

Figure 2 shows the curves for one sequence of "*happy*" characters formed by different people. It can be seen that the curves of one person do not match, and also large differences are displayed for different people. In addition, the writing element can form lines of different thicknesses, which affects the formation of curves (Fig. 2).

The use of sequential enumeration of the coordinates of movement of the writing element does not give a high accuracy of identification of a person, since the size and shape of the characters change with each new formation of the character. Coordinate sequences have different sets of numbers, even if the symbols form the same person. In addition, each generated coordinate sequence is tied to temporary values.

Another more effective method is to form a sequence consisting of symbols of the directions of movement of the writing element. Also recorded are time points for each direction of movement. According to this approach, a code of directions in time is generated, which repeats the shape of a symbol or group of symbols.

For each method, a database of standards is formed, which determines the input sequence by comparison. An already known character sequence or single character is also used. Can also be used a sequence of formation of a geometric figure.

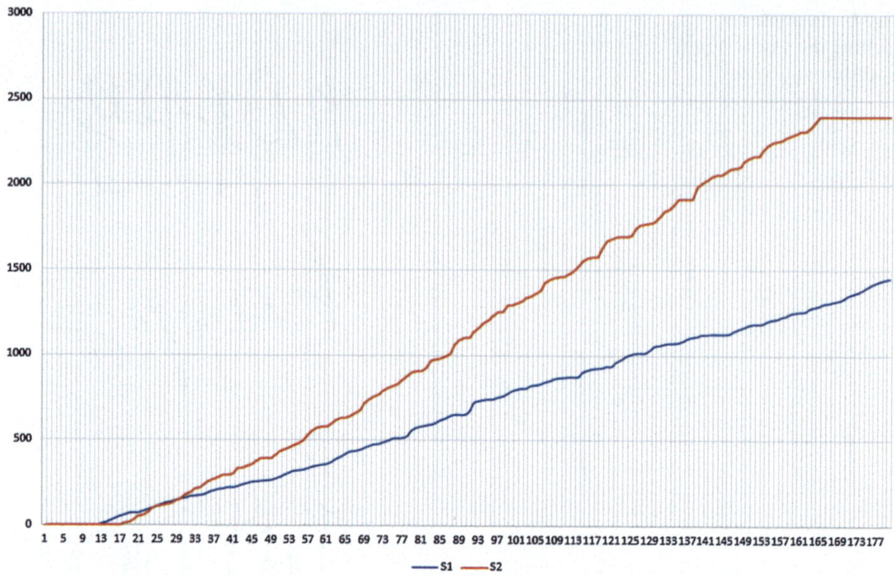


**Fig. 2** Dependences of the number of pixels forming handwritten text on each time frame of a video sequence

Identification can be carried out by studying the entire sequence as a whole, or by analyzing individual elements of the forming handwritten text. Moreover, each individual element of the sequence is selected using special methods. In this case, the sequence element may not be a separate symbol, but elements forming a separate symbol. Fragments of transitions from symbol to symbol can also be used.

Handwritten text can be formed and read by the system by moving one point along an individual path that is not displayed and not fixed, and also text can be formed with a complete sequential display and fixation of the sequence. As a writing element and a reading device, special devices with a contacting surface can be used, which can have different physical realizations (electrical reading, optical reading, and others). An ordinary writing pen (pencil) and paper can be used. It can also use a computer mouse, which has the most convenient form for the free formation of handwritten text.

A special device can generate the necessary signal sequences, and a video sequence or a code of a sequence of numbers that can be stored and analyzed can also be generated. In this paper, handwritten text is analyzed in two versions:

- The video sequence is formed and the frames of the formed video sequence are analyzed;
- Analysis of the dynamics of the formation of handwritten text based on cellular automata (CA).

The main problem is the possibility of falsification of the generated numerical sequence, which displays the dynamics of the formation of handwritten text.

However, such a sequence can be introduced into the system without the use of a writing element and a reading surface, which is usually quite difficult to implement. The movement of a writing element in time is now not difficult to implement. Difficulties arise when reproducing microaccelerations of movements and micro-slowdowns of movements of a writing element. Therefore, the use of characteristics of changes in speed of movements is the most reliable approach to identifying a person by the dynamics of handwritten text.

## 3 Biometric Identification Method Based on the Analysis of the Elements of the Dynamics of Handwritten Text

The methods discussed above use the analysis of movements and reproduced forms at each time step. In fact, standard time intervals are used, on each of which it is determined what is formed and how is formed. To implement the methods at each time step, it was determined how many pixels were added and how they are located on the surface of the image carrier. Also used is the definition of time intervals during which selected elements of the geometric forms of handwritten text are formed.

Based on the obtained quantitative values, average values and confidence intervals were formed for each set value of each characteristic attribute. A threshold percentage is selected, the excess of which gives the identification of “own”. If the quantitative values of characteristic features do not fall within the boundaries of confidence intervals, then the person is identified as an “impostor”. If the percentage of matches does not exceed the selected threshold percentage value, then the identity at the input is defined as an “impostor”. The threshold value is determined as a result of experimental studies.

The method is aimed at analyzing the dynamics of the formation of handwritten text using the “mouse”. Handwritten text is formed using a graphical editor (for example, paint.net) on a white canvas in black with the selected thickness of the writing element (1 pixel). At the same time, using a special program, video is generated from the screen, which is represented by an AVI file. This video file is divided into a sequence of frames, which are separate files in a graphic format (Fig. 3). Video was formed using an application program Camtasia Studio 7.

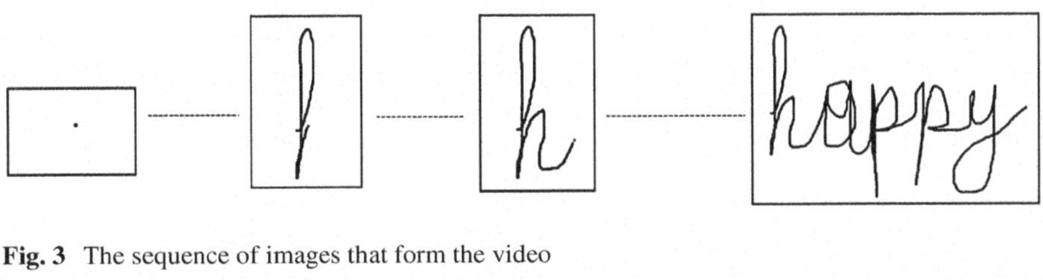


**Fig. 3** The sequence of images that form the video

The graphic files of the video sequence are used to analyze the changes in the images of the generated handwritten text. As a result of the analysis of the images of the video sequence, the formation time of the selected element of the handwritten text is determined. To do this, the number of frames spent on the formation of the image element is determined and multiplied by the time of formation of one frame of the video sequence. Also, the analysis of the video sequence allowed us to determine the number of pixels forming the handwritten image elements.

The third important parameter that is determined is the direction of movement of the writing element. The direction of movement of the writing element is determined in each frame by analyzing the projections of the added part of the image on each frame of the video sequence. To do this, it is necessary to determine and select the added part of the image on each video frame.

Use of the software that allows to generate an additional sequence of frames that contain pixels formed at a given time step (Fig. 4).

Pixels that have changed their state are selected in red in each frame. In the bottom row are excerpts of images in which the selected pixels are enlarged for the corresponding video frames. The selection of pixels in each subsequent frame is based on the method of analysis of neighboring frames, according to which the difference in pixel codes for two frames is determined. If the difference is non-zero, then on the next video frame this pixel is highlighted. Thus, those pixels on the next video frame that have changed their code compared to the same pixels on the previous frame are highlighted.

As a result, the number of pixels added on each video frame using a moving writing element is determined. The selected pixels in each video frame have a specific shape. If the number of selected pixels is large, then this indicates a high speed of movement of the writing element. However, there are frames in which there are no selected pixels, which means there is no movement, or the speed of the writing element is much less than the speed of formation of the frames of the video sequence.

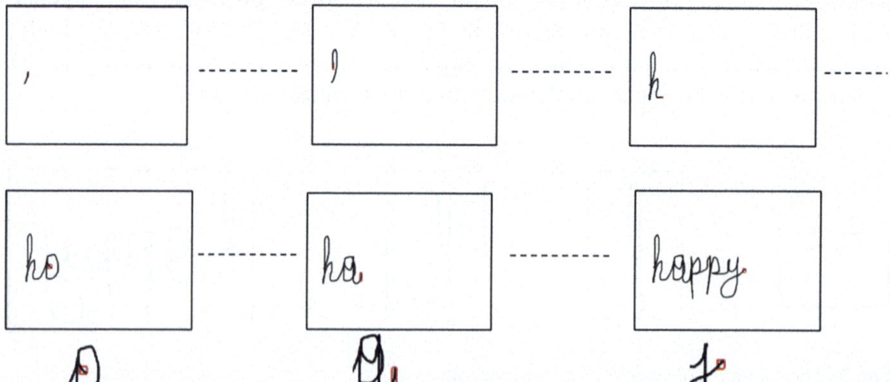


**Fig. 4** An example of forming an additional video sequence with selected pixels

When an additional sequence of video frames is formed, the direction of movement in each video frame is determined. For this, projections of the figure formed by the selected pixels in each video frame are formed. If to use an orthogonal coating, then can clearly form two projections horizontally and vertically. Projections formed from different angles require additional calculations. The number of generated projections determines the accuracy of determining the direction. The more projections, the more accurately the direction of movement of the writing elements is determined.

In this work, a hexagonal coating is used based on studies conducted in [9–13]. The use of hexagonal coating and the Radon transform makes it possible to obtain six projections of the Radon transform, which are formed in the directions $0^0$, $30^0$, $60^0$, $90^0$, $120^0$, and $150^0$.

Each projection is analyzed and the projection length of the selected object is determined. The projection length is calculated as the number of pixels between two extreme non-zero pixels. For example, the projection represented by code 0010111011100 corresponds to nine pixels. Among all the formed projections, the projection that has the smallest length is selected. The direction in which this projection was formed is determined. This direction is the direction of movement of the writing element at the corresponding time step.

When determining the direction of movement, several situations arise.

1. Defined one projection with the smallest length.
2. Two adjacent projections with the smallest length are defined.
3. More than two projections with the smallest length are defined.
4. Not defined single projections.

In the first situation, the direction of movement is determined by the minimum projection. In the second situation, a direction with a smaller angle is determined. In this case, the direction can be defined as the average direction between two minimal projections. For example, if minimal projections are present in directions $30^0$ and $60^0$, then direction $45^0$ is selected. The third situation appears when there is movement by one pixel or by a group of pixels, which is equal to the diameter of the writing element. In the third situation, the direction is not determined.

In the fourth situation, the direction is also not determined. However, in this situation, as in the previous (third) movement of the writing element is possible, since the trajectory can be superimposed on a previously formed handwritten image element.

Given the described limitations and selected characteristic features, sequences of numbers are obtained that determine the area of the generated image of the handwritten text and the numerical values of the directions of movement at each time step. Examples of graphical representations of such sequences on Figs. 5 and 6 are presented.

Charts have significant differences for different people. However, they have many coincidences for several video sequences formed by the same person. It can be seen that coincidences are present on many video frames (Fig. 6). At the same time, there are elements of curves that coincide in shape, but are formed on different

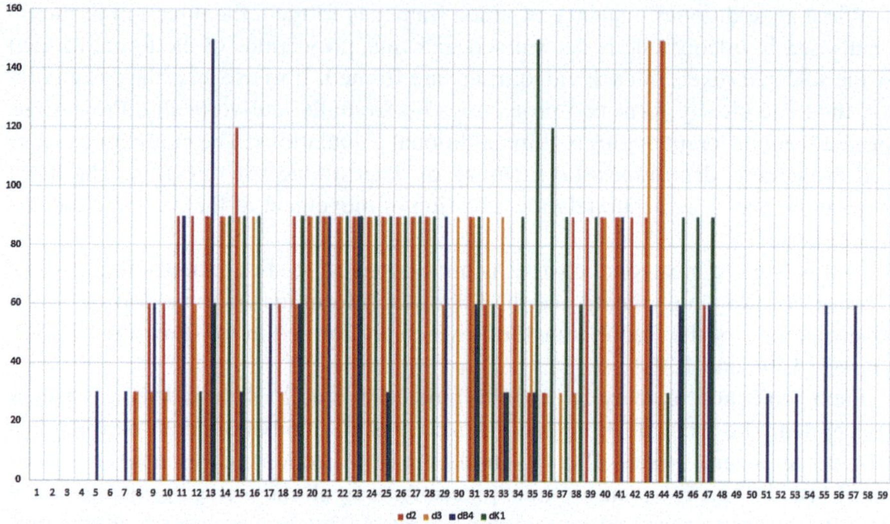


**Fig. 5** Graphs of dependencies of directions for one person (2 graphics, orange and red) and several people (2 + 2 graphics, additional blue and green graphics)

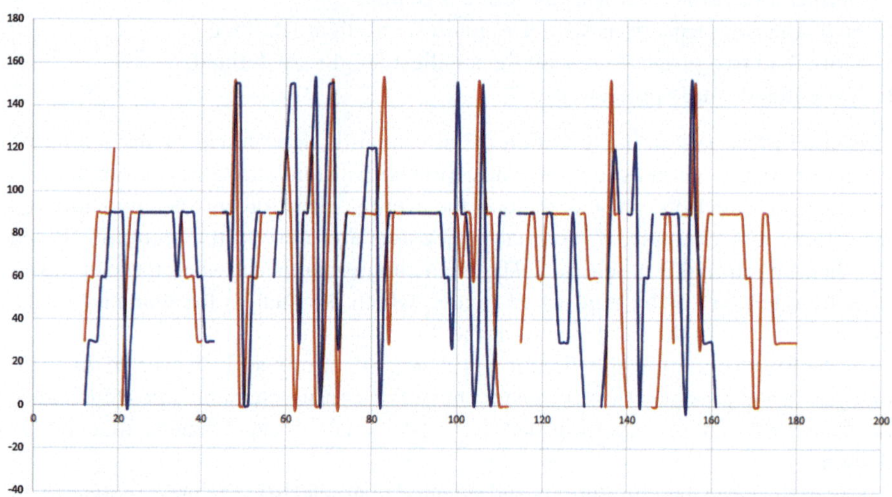


**Fig. 6** Graphs of dependencies of directions for one person

frames of video sequences. It takes into account that identical forms are located on the closest frames of both video sequences. A search is made for such sequence elements by shifting them vertically and horizontally. In the presented example, the number of matches increases as a result of additional shifts.

The described method showed good results. However, the application requires preliminary user training, and also does not take into account all directions of movement in the range from $0^0$ to $360^0$. Also, the method loses information at the points of intersection of the trajectories of the handwritten text.

## 4 Using Cellular Automata to Analyze Handwriting Dynamics

Using the CA allows you to take into account all directions in the range from $0^0$ to $360^0$, as well as analyze the direction of movement at the intersection of the text paths [14, 15]. The handwritten text in dynamics is applied to the cells of the CA. From the moment the trajectories of the handwritten text of the last first cell begin to form, the symbol belonging to the image begins to fulfill the local transition function and puts it in a logical "0" or active state. The zeroed or active cell forms the beginning of the direction, and the next zeroed or active cell forms the end of the first direction and the beginning of the second direction.

At the same time, the number of formed cells is fixed, which determines the distance between the formed direction and the current cell of the formed path.

Directions are formed according to the accepted coding for different types of neighborhood (Fig. 7).

Each number determines the direction of reproduction of the path. In this case, only changes in directions are recorded. Training is carried out by repeatedly forming the symbol trajectory, taking into account local distortions. Local curvatures are taken into account using a large neighborhood. A second-order neighborhood (Fig. 7) is used to determine local distortions that are present in a first-order neighborhood.

During the analysis, neighboring cells that belong to the trajectory are determined. Their quantity and neighborhood are determined. Those ones cells that do

**First Order Neighborhood**

| | | |
|---|---|---|
| 1 | 2 | 3 |
| 8 | 0 | 4 |
| 7 | 6 | 5 |

**Second Order Neighborhood**

| | | | | |
|---|---|---|---|---|
| 9 | 10 | 11 | 12 | 13 |
| 24 | 1 | 2 | 3 | 14 |
| 23 | 8 | 0 | 4 | 15 |
| 22 | 7 | 6 | 5 | 16 |
| 21 | 20 | 19 | 18 | 17 |

**Fig. 7** Coding of directions in CA

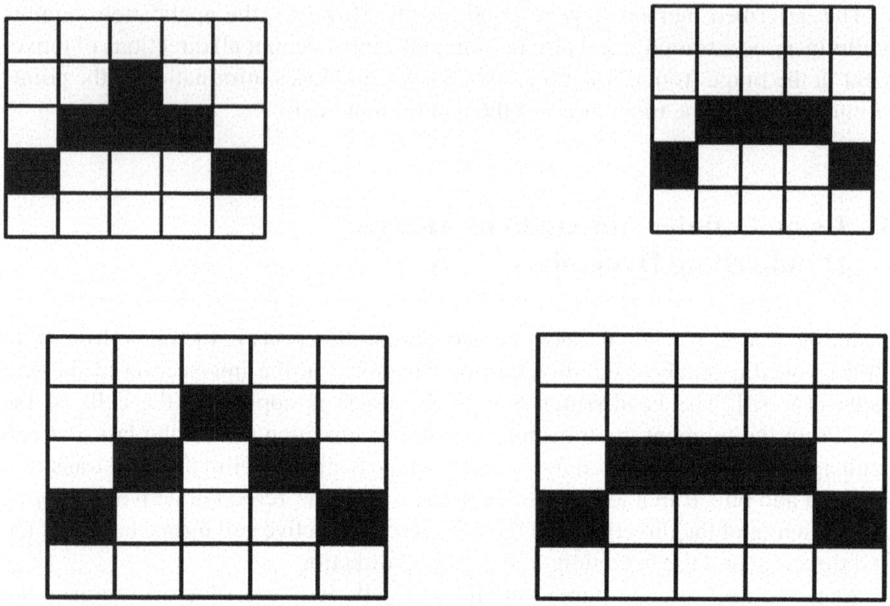

**Fig. 8** Examples of normalizing the trajectory of a curve of handwritten characters

not give the shortest distance to the cells of a second-order neighborhood that belong to the trajectory are zeroed. Examples of such situations are considered in Fig. 8.

This procedure is implemented after the operation of thinning the trajectory. The trajectory in time is fixed using special trajectory tables (Table 1) or graphical histograms of the directions of movement (Fig. 9).

According to the shape of the graph, a symbol is determined. The sequence of the location of each quantity is important. The sequence of each quantity is determined, and then the duration of formation of each quantity is determined, which are compared with the average reference value. Scale factors are taken into account, and if in a larger number they do not change the shape of the graph, the symbol is identified.

For experimental studies of this method, the FinSimbols software was used, which allows one to form a symbol on the cell medium and normalize its trajectory. The program also implements the formation of a reference database with which identification is carried out (Fig. 10).

The program implements the following algorithms. All algorithms initially perform the same steps.

A handwritten character is entered into the cell environment. In the course of the dynamics of input, a code of directions is formed, which is recognizable. Before recognition, training is conducted, which consists in forming a base of reference codes.

The list of reference codes begins in the recognition process. If there is at least one code in the reference that matches each other with the code at the input, then the algorithm terminates and shows a character corresponding to the code found.

**Table 1** Trajectory table for symbol «*a*»

| | 1 | 2 | 3 | 4 | 5 | 6 | 7 | 8 | Code |
|---|---|---|---|---|---|---|---|---|---|
| 1. | | | | | | | | | 1(2) |
| 2. | | | | | | | | | |
| 3. | | | | | | | | | 8(2) |
| 4. | | | | | | | | | |
| 5. | | | | | | | | | 7(2) |
| 6. | | | | | | | | | |
| 7. | | | | | | | | | 6(4) |
| 8. | | | | | | | | | |
| 9. | | | | | | | | | |
| 10. | | | | | | | | | |
| 11. | | | | | | | | | 5(2) |
| 12. | | | | | | | | | |
| 13. | | | | | | | | | 4(2) |
| 14. | | | | | | | | | |
| 15. | | | | | | | | | 3(2) |
| 16. | | | | | | | | | |
| 17. | | | | | | | | | 2(4) |
| 18. | | | | | | | | | |
| 19. | | | | | | | | | |
| 20. | | | | | | | | | |

Otherwise, the program compares the input code sequences and the base reference codes. If the sequence is the same, the position of this code is stored in the table.

If the sequence does not match, then cyclical shifts of all digits of the code begin. The first character of the code is placed last and the sequence is compared with the sequence of the reference code. The next match stores the position of this code in the table. Otherwise, the offset is carried out until the input code is reset.

Then the same operations are performed, but the reference code is already shifted. After that a cycle from 1 to 8 is carried out, and in the code at the input "1" is added to each value (if the value is 8, then instead of adding a unit the value is replaced by "1") and in each pass of the cycle the sequence of the shifted code is compared with the reference one. If the sequence is the same, then the position of this code is stored in the table.

When these operations are completed, different steps begin for each method.

For algorithm 1.

All reference codes in which the sequences are matched are checked. The number of values "1", "2", …, "8" is calculated and one by one, the received sums are written to the array [1 … 8]. The same thing happens with the code at the entrance. Then in parallel these numbers are divided and the relations are summed up as a result. After that the reverse division is made and the relations are also summed up. The amount that is less and saved is chosen. The code in which this amount is less and is considered the closest.

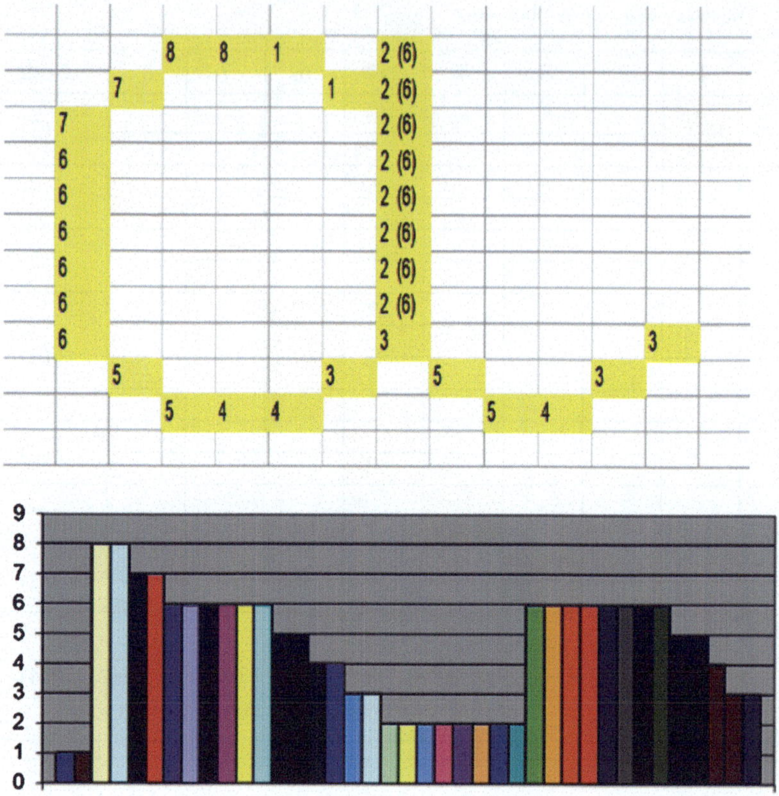


**Fig. 9** An example of the distribution of the directions of movement of a writing element for a symbol «*a*»

For algorithm 2.

All reference codes in which the sequences are matched are checked. The number of values "1", "2", ..., "8" is calculated and the amounts are written in the order in which the values go in the code. Then the sums of the first group are divided by the second, the second by the third, etc., and the relations themselves are summed up by one sum. The same operations are performed with the code at the entrance. The reference code with the closest sum of relations to the sum of the relations of the code at the input is considered the closest.

For algorithm 3.

A reference code that has the closest difference between the origin and end coordinates of the input code is the closest.

For algorithm 4.

One and double errors are corrected. If no code is found among the reference codes, the sequence of which is the same as the code at the input, then one character is deleted alternately from the code at the input. The same operation is performed in each position. All this time its sequence is compared with the sequence of reference

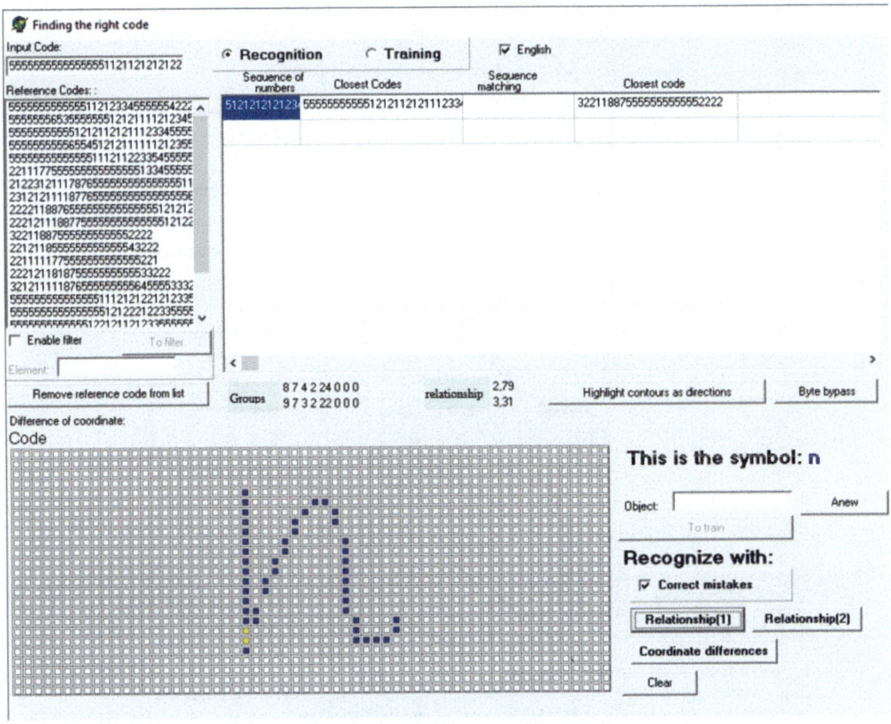


**Fig. 10** Handwritten character identification program interface

codes. If the sequence is the same, then the function terminates and stores the position of this code in the table. Otherwise, the same thing happens. In this case, two characters in all possible combinations are sequentially deleted from the input code and the sequence is compared again.

Conducting the experiment consisted in the formation of standards of symbols "*0*" and "*C*." Recognition performance was determined at different standards and was given by the ratio of the number of recognized images to the total number of identifiable images.

The "*O*" and "*C*" symbols were recognized, taking into account large-scale changes with and without orientation change. The presence of single and double errors was taken into account, as well as the different beginning of the character drawing. During the experiment, it was taken into account that a complete search of all combinations is not possible because recognition is applied to characters in scale and orientation changes.

For each symbol the image forming 30 times were performed and the probability of true recognition, false recognition and the probability of uncertainty of recognition were determined. Different variants of characters and their recognition by the program are presented in Fig. 11.

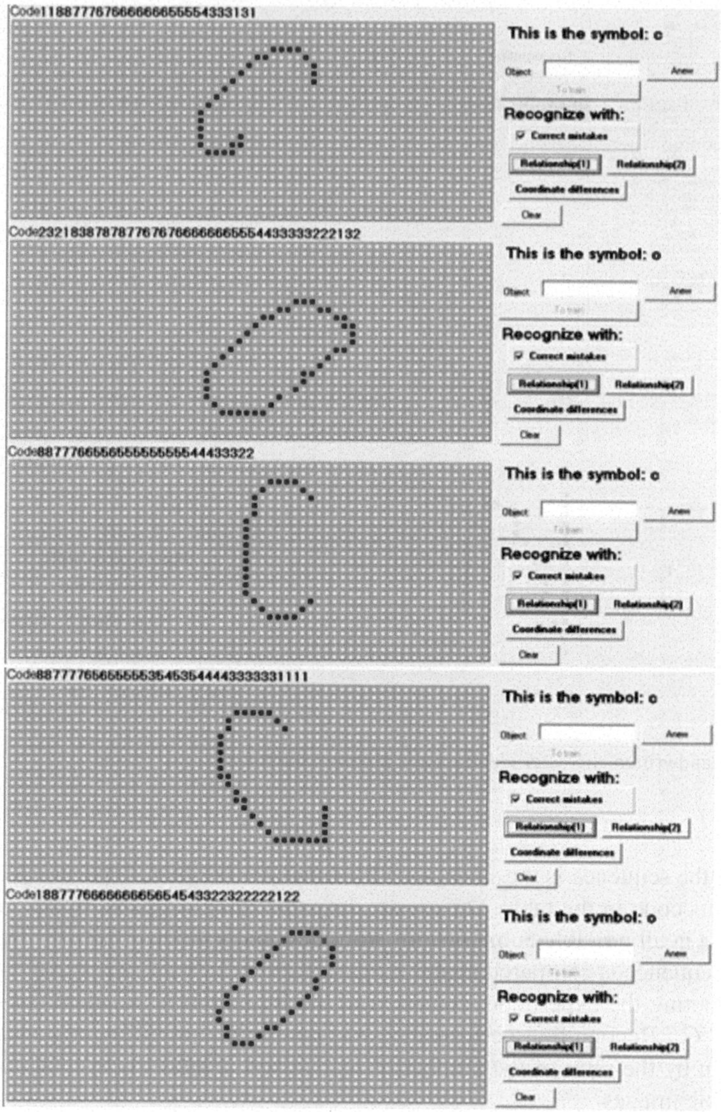


**Fig. 11** Program options for recognizing “O” and “C” characters with different number of standards

# 5 Conclusion

Personality identification by the dynamics of handwritten text is increasingly used in various fields of human activity. Compared with identification systems for static images of handwritten text, algorithms that realize identification by the dynamics of handwritten text are simpler. The implementation of algorithms based on the analysis of areas and directions allows to reduce the number of computational operations, as well as to analyze the elements of the formed sequence using simple shift operations. Representation of the dynamics of the formation of a handwritten text increases the number of matches of quantitative values of characteristic signs with reference values due to the implementation of shifts of a group of sequence elements in the time scale and direction scale values. Using cellular automata determines the direction of movement of the writing element in the range from $0^0$ to $360^0$. Also, cellular automata display the direction of movement of the writing element at the intersection of the trajectories of the curves when the symbol is displayed in time. The developed software allows you to adapt these methods to various means of entering graphic information, as well as form a database of reference sequences for the implementation of the identification process.

In the future, the authors plan to conduct research in the field of biometric identification on the dynamics of writing a handwritten text by a person using various devices for graphic input of information into the system.

## References


1. Accardo, A. P., Genna, M., & Borean, M. (2013). Development, maturation and learning influence on handwriting kinematics. *Human Movement Science, 32*(1), 136–146.
2. Alamargot, D., & Morin, M.-F. (2015). Does handwriting on a tablet screen affect students graphomotor execution? A comparison between grades two and nine. *Human Movement Science, 44*, 32–41.
3. Asselborn, T., Gargot, T., Kidzínski, L., Johal, W., Cohen, D., Jolly, C., & Dil-lenbourg, P. (2018). Automated human-level diagnosis of dysgraphia using a consumer tablet. *NPJ Digital Medicine, 1*(1), 42.
4. Lozhnikov, P., & Chernikova, O. (2011). Handwriting dynamics as a means of authentication. In *6th International Conference on Internet Technology and Secured Transactions, 11–14 December 2011* (pp. 176–179).
5. Rémi, C., Prévost, L., & Anquetil, E. (2015). Drawing, Handwriting Processing Analysis: New Advances and Challenges. In *Proceedings of the 17th Biennial Conference of the International Graphonomics Society Pointe-à-Pitre & Saint-Claude, Guadeloupe, June 21-25*.
6. Harralson, H. H., & Miller, L. S. (2017). *Huber and Headrick's handwriting identification: facts and fundamentals* (2nd ed.). Cleveland, OH: CRC Press.
7. Rosa, G. H., Papa, J. P., & Scheirer, W. J. (2018). Person identification using handwriting dynamics and convolutional neural networks. *Deep Learning in Biometrics*, 227–244.
8. Huber, R., & Headrick, A. (1999). *Handwriting identification: facts and fundamentals*. Boca Roton: CRC Press.
9. Motornyuk, R. L., & Bilan, S. (2019). Methods for extracting the skeleton of an image based on cellular automata with a hexagonal coating form and radon transform. In S. M. Bilan & S. I. Al-Zoubi (Eds.), *Handbook of Research on Intelligent Data Processing and Information Security Systems* (pp. 289–329). Hershey, PA: IGI Global.
10. Motornyuk, R. L., & Bilan, S. (2019). The moving object detection and research effects of noise on images based on cellular automata with a hexagonal coating form and radon transform. In S. M. Bilan & S. I. Al-Zoubi (Eds.), *Handbook of Research on Intelligent Data Processing and Information Security Systems* (pp. 330–359). Hershey, PA: IGI Global.
11. Bilan, S., Motornyuk, R., & Bilan, S. (2014). Method of hardware selection of characteristic features based on radon transformation and not sensitive to rotation, shifting and scale of the input images. *Advances in Image and Video Processing, 2*(4), 12–23.

12. Belan, S. N., & Motornyuk, R. L. (2013). Extraction of characteristic features of images with the help of the radon transform and its hardware implementation in terms of cellular automata. *Cybernetics and Systems Analysis, 49*(1), 7–14.
13. Motornyuk, R.L. (2013). Computer-aided methods for identifying images of moving objects based on cellular automata with a hexagonal coating. Dissertation for the degree of candidate of technical sciences (UDC 004.932: 519.713 (043.3)), Kiev: SUIT, 2013.
14. Wolfram, S. (2002). *A new kind of science*. Champaign, IL: Wolfram Media.
15. Bilan, S. (2017). *Formation methods, models, and hardware implementation of pseudorandom number generators: Emerging research and opportunities*. Hershey, PA: IGI Global.

# Gait Identification Based on Parallel Shift Technology

**Stepan Bilan, Ruslan Motornyuk, Sergey Yuzhakov, and Mariia Halushko**

## 1 Introduction

Modern population growth requires the use and improvement of methods and means of biometric identification. With the increase in the number of friends and acquaintances, it is becoming more and more difficult for each person to remember each of them. Especially there are many cases when one person has to interact with a large number of people. For example, a teacher gives lectures and conducts practical and laboratory studies with a large number of students and he cannot individually remember each of them. In access systems, also a person is not able to identify everyone who requests access. In this regard, biometric identification tools are indispensable.

There are many biometric characteristics that cannot be represented by a single measured quantity. As a rule, all biometric characteristics are determined by a multitude of quantitative values. These quantitative values occupy a certain place in the set of quantities that allows you to identify a person.

Human biometric characteristics are divided into static and dynamic [1–3]. Static biometric characteristics are determined in close proximity to a person or at a predetermined beforehand distance from a biometric detector. For example, fingerprint, face, hand, ear shape, finger vein pattern, iris, and other biometric characteristics cannot be determined at arbitrary distances from biometric scanners. Many of them

S. Bilan (✉) · M. Halushko
State University of Infrastructure and Technology, Kyiv, Ukraine
e-mail: bilan_sm@gsuite.duit.edu.ua; mariiahalushko@meta.ua

R. Motornyuk
PU “Kiev Department” Branch of the Main Information and Computing Center of the JSC “Ukrzaliznytsya”, Kyiv, Ukraine
e-mail: xehap0@sw.uz.gov.ua

S. Yuzhakov
Haisyn Department of the SFS General Directorate in Vinnytsia Region, Vinnytsia, Ukraine
e-mail: yserg74@meta.ua



S. Bilan et al. (eds.), *Biometric Identification Technologies Based on Modern Data Mining Methods*, https://doi.org/10.1007/978-3-030-48378-4_5

require very small distances. For example, a fingerprint can be read by directly touching the scanner with a finger.

However, there are situations where static biometric characteristics cannot be determined due to large distances, noise, and other reasons. Then a search is made for such biometric characteristics that would make it possible to identify a person. One of these characteristics is gait. This characteristic does not require strictly fixed distances and high-precision measuring instruments. A person's gait is analyzed for a certain period of time, and the result is a study of the dynamics of changes in quantitative characteristics over time. For each person, such dynamics is individual. There are many tasks in which to identify a person there is only data about his gait based on a fixed silhouette without displaying other biometric characteristics. Such tasks arise when shooting videos in difficult conditions (night time, high dustiness, poor characteristics of video cameras, etc.).

To date, the gait has been little studied and there are no high-precision methods and means of biometric identification for this parameter [1, 4–9]. In this direction, modern developments are actually marking one place in development and improvement. Therefore, for many reasons, the task of biometric identification of a person by the dynamics of his gait is one of the first tasks in many automated systems.

This chapter solves the problem of biometric identification of a person by the dynamics of his gait by analyzing a video sequence using the theory of cellular automata and parallel shift technology.

## 2 Formulation of the Problem and Relative Works

Throughout the world, researchers are actively working on the biometric identification of a person by his gait. Various methods are used to solve this problem [4–9]. The most popular methods are those based on the analysis of a video sequence, the analysis of the shadow of a moving person, and also using various sensors mounted both on different parts of the human body and on the surface on which it moves [10]. Great successes have already been achieved in these areas, which made it possible to determine the main characteristic features in the analysis and classification of human gait. The use of sensors mounted on the surface and in various parts of the human body limits the scope of such methods, since the person must be identified in a specific place. Sensors make this kind of biometrics addictive.

Human gait becomes an independent type of biometrics when using cameras [1, 5, 6, 11]. With the help of video cameras, a walking person can be fixed at any distance and at any angle. Moreover, for each angle and distance, various methods of data transformation and the formation of characteristic features are used.

In addition, gait is a complex characteristic, as it can change for the same person. For example, the gait of one person changes when walking on various surfaces (asphalt, grass, sand, etc.). Also, for different angles of video cameras, the shape of the human body changes on each frame of the video sequence. The main characteristic of a person's gait is the repetition period of his movements. At the same time, the duration of the repetition of movements can also change if a person starts to walk faster or slower, as well as make a jerk, etc. In the analysis of gait, there are still many problems associated with the instability of human movement. However, in certain areas of movement, a person goes into a stable mode, the analysis of which allows to identify a person. Modern methods are aimed at analyzing the shape of the body of a moving person, as well as determining the shape of a person in each video frame to determine the trajectory of his movement and the movement itself. These methods make calculations and intelligent operations more complex and use large amounts of data for analysis.

In connection with existing problems, the task is to create methods and means that in any variant of the person's location identified him regardless of the quality presentation of the video data.

## 3 Theoretical Foundations of Parallel Shift Technology

The current state of the issue of image recognition has formed a number of unsolved problems in this area. The main problem is not solved—the automatic formation of characteristic features for all image classes.

Today there are many artificial systems for processing and recognizing images of specific classes (images of text, faces, fingerprints, identifiers, etc.). Each image from such classes is recognized using a set of characteristic features, which the recognition system developer has previously formed. The developer preliminary analyzes the images of the selected class and, based on his own experience, selects a set of characteristic features. The developer develops methods and tools for determining the quantitative values of the proposed characteristic features. Image recognition processes of the selected class in Fig. 1 are presented.

This diagram does not show the processes of learning and finding the nearest reference. The methods for finding the nearest standard have been developed. However, their implementation requires a large investment of time and hardware. Also, the training mode of the system that is used to form the reference memory is not indicated.

The mechanism for recognizing an object in an image with different forms of its representation was not found. A set of characteristic features must be formed so that

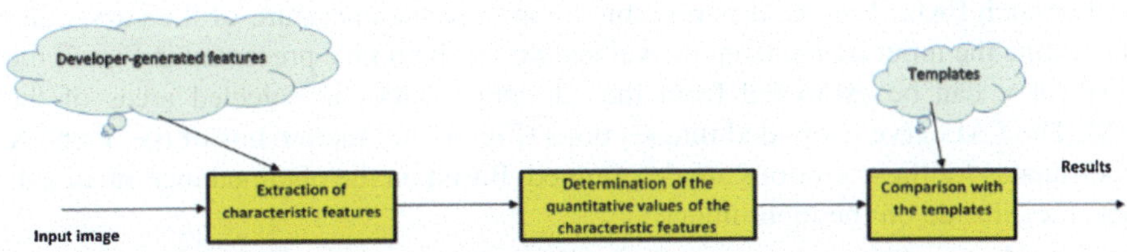


**Fig. 1** Image recognition processes for one class

it can describe the image as a whole. A characteristic feature is a quantitative characteristic that displays information about one of the image properties.

The formation of quantitative characteristics requires complex measurements or calculations. In this case, measurements and calculations are used. This situation complicates the methods of preliminary image processing and the selection of characteristic features.

To simplify the methods for distinguishing characteristic features, it is necessary to reduce the number of complex measurements and calculations. For this, it is necessary to reduce the number of characteristic features. If you reduce the number of characteristic features, then the developer must choose features that allow you to describe all the properties of the image. The task is to search for image transformations that make it possible to describe all image properties using selected characteristic features.

The paper proposes to reduce the number of characteristic features, thereby reducing the number of computational operations. This approach presents an image in the form of a set of quantitative values of one characteristic feature, which fully describes all the properties of images.

To solve this problem, it is proposed to use the technology of parallel shift (PST) [12–14]. This technology is based on calculating the intersection area of the original image with the image of its copy, which is shifted in different directions. The effectiveness of this technology has been proven experimentally, and the results are presented in many publications.

The main characteristic of PST is the function of the area of intersection (FAI), which is defined as the dependence of the intersection area of the figure and its copy, which is parallelly shifted in the direction φ at each iteration step (Fig. 2).

Only four shift directions are used ($0^0$, $45^0$, $90^0$, and $135^0$), since the other directions have the same FAI values as the four directions presented. Other shift directions (for example, $30^0$, $60^0$, $120^0$, etc.) can also be used, which do not repeat in the range from $0^0$ to $180^0$. As a rule, directions are used that are conveniently implemented on orthogonal coverage and practically do not give image distortions. The FAIs shown in Fig. 2 describe geometric shapes with a small number of pixels and a large number of differences at the intersection of areas. FAI curves for such images have a complex structure. Therefore, the analysis is carried out in one way, which consists in analyzing the quantitative characteristics of the obtained numerical sequences. Extreme function values are analyzed and the input image is determined.

In most cases, there are images consisting of a large number of pixels and have an FAI with a more even characteristic (Fig. 3).

For such FAIs (Fig. 3), it is possible to analyze the curvature of the curve, and determine the input image from the values. So for the image presented in Fig. 3, the FAI curve can be estimated from the curvature values in selected areas of the FAI. The FAI curve is divided into sections (Fig. 4) and increments of the Y and X coordinates for these sections are determined. Based on the obtained increment values, they decide on the input image.

**Fig. 2** An example of FAI formation for shift directions $0^0$, $45^0$, $90^0$, and $135^0$

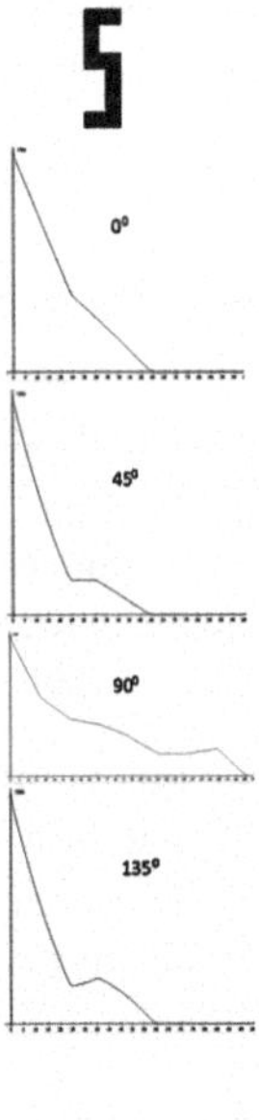


**Fig. 3** FAI example for high pixel images

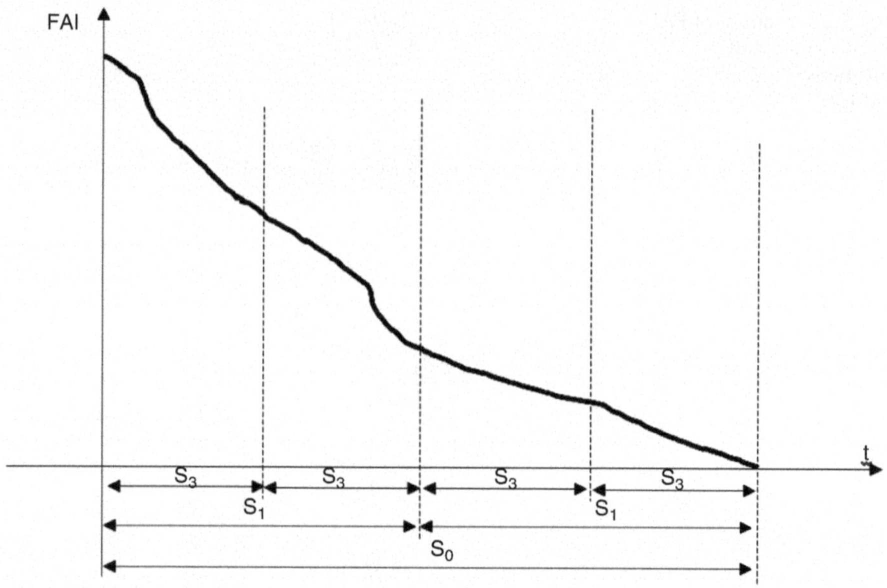


**Fig. 4** An example of splitting an FAI curve into sections

For the described example (Fig. 3), the increment values for each shear direction are presented in Fig. 5. The larger of the divided intervals, the more accurately you can describe the image at the input.

This technology is very successfully used for problems of image analysis and recognition using one characteristic feature—the intersection area. This approach reduces the number of operations and the algorithms used, so only two operations are performed: shift and area calculation. These operations do not require large computational resources.

## 4 Method Description

The method of analyzing a person's gait and its biometric identification is based on processing of images of a moving person created using conventional, infrared (thermal), and other video cameras. The gait of a person is analyzed, represented by cameras located at any angle to a moving person.

The method consists of several stages.

1. Image binarization.
2. The moving object selection in an image using a descriptive rectangle.
3. The selection of pixels (cells) that in the image of each frame have changed their properties compared to the previous frame of the video sequence.

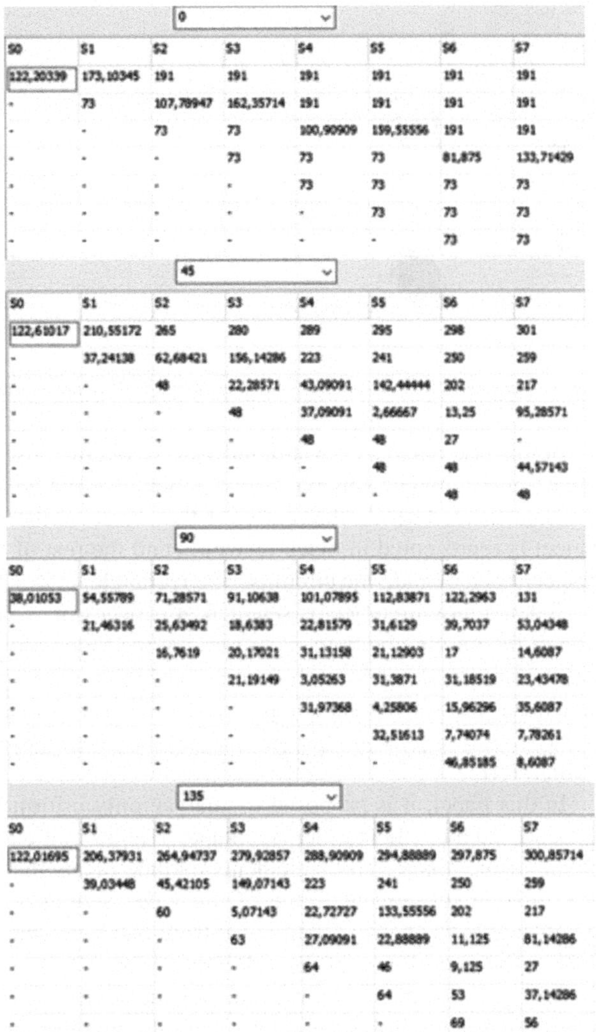

0

| S0 | S1 | S2 | S3 | S4 | S5 | S6 | S7 |
|---|---|---|---|---|---|---|---|
| 122,20339 | 173,10345 | 191 | 191 | 191 | 191 | 191 | 191 |
| - | 73 | 107,78947 | 162,35714 | 191 | 191 | 191 | 191 |
| - | - | 73 | 73 | 100,90909 | 159,55556 | 191 | 191 |
| - | - | - | 73 | 73 | 73 | 81,875 | 133,71429 |
| - | - | - | - | 73 | 73 | 73 | 73 |
| - | - | - | - | - | 73 | 73 | 73 |
| - | - | - | - | - | - | 73 | 73 |

45

| S0 | S1 | S2 | S3 | S4 | S5 | S6 | S7 |
|---|---|---|---|---|---|---|---|
| 122,61017 | 210,55172 | 265 | 280 | 289 | 295 | 298 | 301 |
| - | 37,24138 | 62,68421 | 156,14286 | 223 | 241 | 250 | 259 |
| - | - | 48 | 22,28571 | 43,09091 | 142,44444 | 202 | 217 |
| - | - | - | 48 | 37,09091 | 2,66667 | 13,25 | 95,28571 |
| - | - | - | - | 48 | 48 | 27 | - |
| - | - | - | - | - | 48 | 48 | 44,57143 |
| - | - | - | - | - | - | 48 | 48 |

90

| S0 | S1 | S2 | S3 | S4 | S5 | S6 | S7 |
|---|---|---|---|---|---|---|---|
| 38,01053 | 54,55789 | 71,28571 | 91,10638 | 101,07895 | 112,83871 | 122,2963 | 131 |
| - | 21,46316 | 25,63492 | 18,6383 | 22,81579 | 31,6129 | 39,7037 | 53,04348 |
| - | - | 16,7619 | 20,17021 | 31,13158 | 21,12903 | 17 | 14,6087 |
| - | - | - | 21,19149 | 3,05263 | 31,3871 | 31,18519 | 23,43478 |
| - | - | - | - | 31,97368 | 4,25806 | 15,96296 | 35,6087 |
| - | - | - | - | - | 32,51613 | 7,74074 | 7,78261 |
| - | - | - | - | - | - | 46,85185 | 8,6087 |

135

| S0 | S1 | S2 | S3 | S4 | S5 | S6 | S7 |
|---|---|---|---|---|---|---|---|
| 122,01695 | 206,37931 | 264,94737 | 279,92857 | 288,90909 | 294,88889 | 297,875 | 300,85714 |
| - | 39,03448 | 45,42105 | 149,07143 | 223 | 241 | 250 | 259 |
| - | - | 60 | 5,07143 | 22,72727 | 133,55556 | 202 | 217 |
| - | - | - | 63 | 27,09091 | 22,88889 | 11,125 | 81,14286 |
| - | - | - | - | 64 | 46 | 9,125 | 27 |
| - | - | - | - | - | 64 | 53 | 37,14286 |
| - | - | - | - | - | - | 69 | 56 |

**Fig. 5** An example of the formation of arrays of increments of coordinates for the curve shown in Fig. 3

4. Analysis of the area of selected pixels in each frame using parallel shift technology and the formation of quantitative data for each frame of the video sequence.
5. Processing the obtained data and forming a quantitative image of a person's gait.
6. Comparison with reference values stored in the database.
7. Decision making about identification.

The first step is to represent the image in binary form. There are tools that form a binary image during filming. For this, thermal, luminescent, and other video cameras are used that respond only to a moving person. To analyze a person's gait, a binary image is enough and there is no need to use a color image.

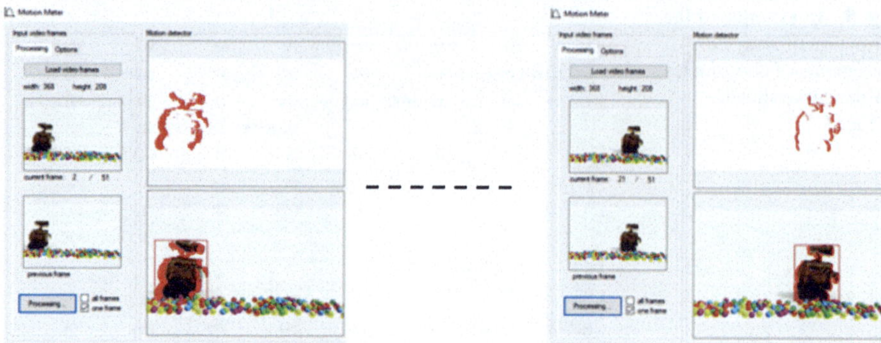

**Fig. 6** An example of selecting a moving object in a visual scene

In the case when a conventional video camera is used, the selection of a moving object in the image using the circumscribe quadrangle is used (Fig. 6). In this case, the data located inside the region of the circumscribe rectangle is read. A moving object is represented in black or white, and the rest of the image is white or black, respectively, outside the circumscribe rectangle.

When a moving object is selected, it is necessary to analyze the shape and location of the moving body in each frame of the video sequence. Since a moving body has a complex shape in each frame after binarization, it is necessary to use special analysis methods that are complex and time-consuming. Therefore, there are methods for analyzing the human gait, focused on the analysis of individual parts of the human form.

In this paper, it is proposed to analyze only informative parts of the image that indicate the movement of a person and changes in the location of his shape on each frame of the image. These elements include pixels that have changed their properties in each subsequent frame of the video sequence. Typically, these pixels are pixels that have changed their color from black to white and from white to black in each subsequent frame of the video sequence.

In Fig. 7 presents several frames of the original video sequence and the corresponding images with selected pixels of the same frames.

As can be seen from Fig. 7, pixels are located on the edges of the shape of a moving object. If the speed of the object is high, a large number of pixels are selected on each subsequent frame. If the speed of the object does not exceed the speed of formation of the next frames, then the pixels are not selected. Thus, the greater the speed the camera has, the more accurately you can identify a person's gait.

Selected pixels make up an image of complex shape, which may be self-color, or may consist of various groups of related pixels. Such an image is difficult to describe and assign a clear identifier to it. The totality of all selected pixels represents the total area of this selected object in each frame. Parallel shift technology (PST) is used to analyze this shape [12–14], which makes it possible to describe any binary image using a set of functions of intersection of areas (FAI). Such functions form a set of numbers that determine the area of the image obtained as a result of the

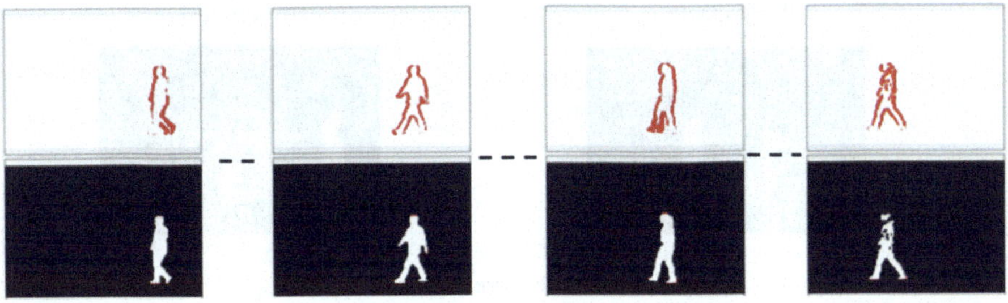

**Fig. 7** An example of the allocation of information pixels in frames of a video sequence

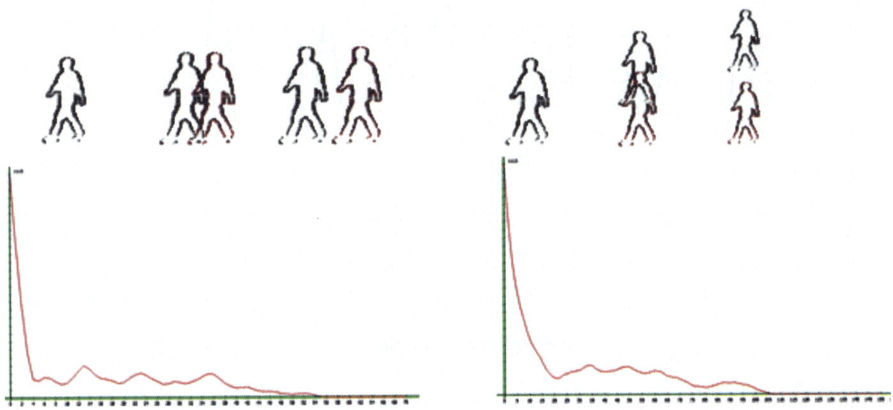

**Fig. 8** An example of the formation of FAI for one frame of a video sequence and for the directions of shift $0^0$ and $270^0$

intersection of the area of the initial image and its copy shifted in one of the directions at each shift step. A set of obtained numbers in several directions of shift uniquely determines the shape of the image. The more shear directions are used, the more accurate the image description.

Thus, for each frame of the video sequence, FAIs are formed, which are then compared to obtain a common relationship that determines the dynamics of movement. First of all, the shape of the curve is determined within the repetition period of gait movements. The length of the gait curve is also determined on the selected number of frames of the video sequence (Fig. 8).

To determine the repetition period of movements, average values for all FAIs obtained on each image of the frame are determined. Thus, the average value of the intersection area on each frame was determined as a result of the shift of the image copy in the frame. The dependence of the average intersection area on the number of frames of the video sequence was built (Fig. 9).

Figure 4 shows the extreme values of the curve, as well as the images of real frames corresponding to these sequences. On these video frames, images of the human body have approximately the same shape. Therefore, the FAI for these video

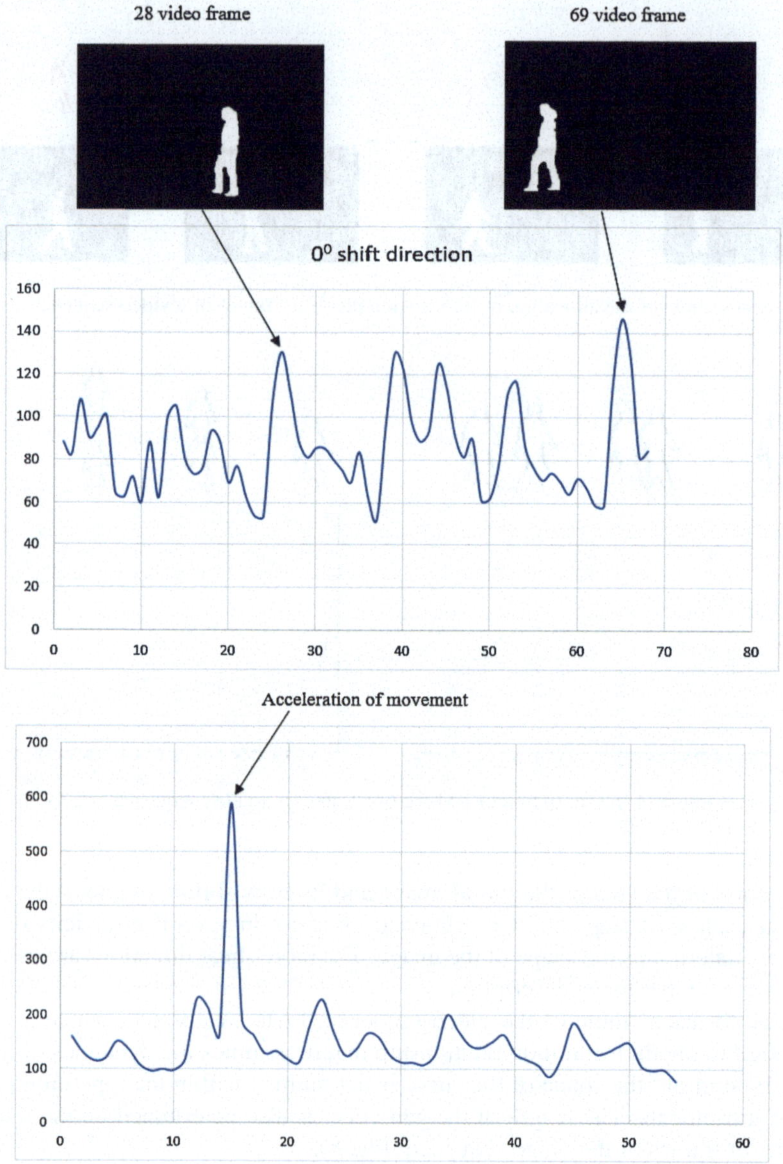


**Fig. 9** An example of the dependence of the average FAI value on the number of frames of a video sequence

frames have the same shape. However, they may vary in scale. The adjacent frames of these extreme frames also have the same shape FAI, as human movements are repeated while walking.

After the gait repetition period is determined, a comparison is made with the reference averages found in the generated reference database.

If the length of the repeat period does not match any reference value, then the gait is not identified. If the repetition period coincides with the reference value within the boundaries of the allowable difference, then the average value curve is analyzed within the boundaries of the period. Typically, a period is defined between several extreme values.

The average value curve is analyzed using the area values, the number of shifts between the maximum and zero area values, the number of vibrations (minimum and maximum extrema). Acceleration during gait is judged by the shape of the extremes. The size of the area determines the approach or removal of a person from the video camera.

Based on the generated FAI values, the average FAI value for each video frame is calculated. As a result, a sequence of average values is formed, which is compared with a sequence of average values located in the reference base. The generated sequences of average values are combined with the maximum extrema of the reference sequences of average values. All numbers of average values on each frame subtract numbers from the corresponding reference sequences. If the obtained $a_i$ values do not go beyond confidence limits $a_i \in [-A; A]$, then for each video frame is formed 1. Otherwise, 0 is formed. The quantity 1 is calculated for a given value A. If, within the repetition period, the number of unit values exceeds a predetermined threshold, then the person is identified as "own," otherwise the walking person is defined as "impostor." An example of identification is presented in Fig. 10.

This identification is used for gait with equal lengths of the gait repetition period.

## 5 Information Pixel Selection Method

In video surveillance systems, a digital video camera generates digitized video on the output, which is come to a cellular automata and to a computer. Thanks to the high speed of the cell environment, the bit mask of pixels in which motion is detected will be transmitted to the computer almost immediately after receiving the image frame. Further, depending on the tasks, a computer program may, for example, frame a moving object, or cut it along that frame from the entire image, or enlarge it to the boundaries of the moving area, etc.

When using an analog video camera, the video stream will only be transmitted to a computer, where it will be digitized and frame-by-frame transmitted to the cellular structure. A cellular automaton selects an array of cells in which movement is detected. This array will be transmitted to the computer for further processing, depending on the tasks that were set for the system.

A cell represents one pixel of a binary image. In the logical state "1," the cell encodes a black pixel, in the state "0"—a white pixel. A simple cell structure is used to work with a binary image. To detect motion in this case, it is necessary to compare the state of the cell in the current and previous frames. In case of coincidence of states ("0" = "0," "1" = "1"), a logical "0" is formed at the outputs of the motion detector. If the states do not match ("0" → "1" or "0" → "1"), then the movement is

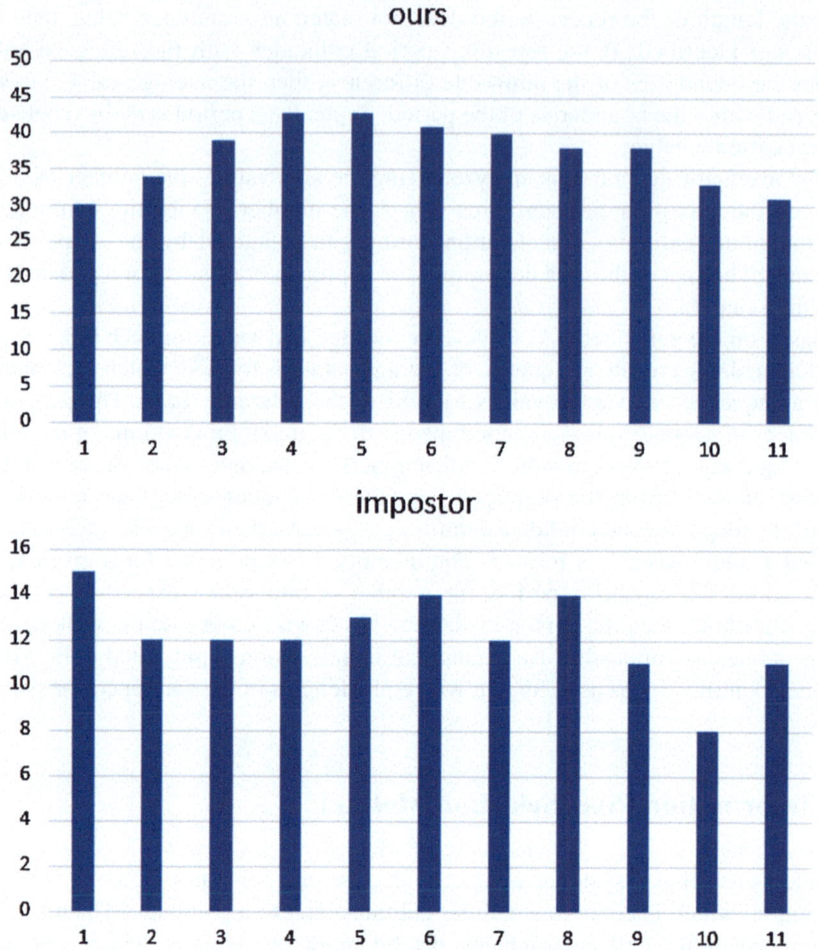


**Fig. 10** An example of the formation of the difference for "ours" and for "impostor"

fixed and the logical "1" is set at the output of the motion detector. For such images, the background is represented in one white color, and moving objects differ from the background in black.

In real conditions, a moving object cannot always clearly differ from the background, which can lead to the merging of low-contrast pixels with background pixels during binarization. Binarization can remove two objects with low contrast. Only objects that contrast strongly with the background cells will remain. The remaining cells clearly selected movement. Binarization gives satisfactory results when working with a contrast image and is generally unacceptable when working with low-contrast real images. To improve quality, it is necessary to use multi-bit cells. The only way out is to use not single-bit, but multi-bit processor elements (PEs), which can encode and compare not only binary, but also multi-gradation images. If you process color images, a lot of redundant information is formed. Therefore, to detect movement it is enough to use a grayscale image. In this case, instead of three RGB channels, one is enough, which at least three times will reduce hardware costs. Decreasing bit depth gives a gain in hardware costs, but too low bit depth negatively affects the quality of allocation of moving objects. The less gradations, the more noise will be destroyed, the less hardware costs for building a PE of cell. But at the same time, more details of the image will be lost and it is more likely that if the contrast between the moving object and the background is insignificant, the object and background will merge. As a result of such a merger, the movement will not be fixed, or not all the moving object will be selected. Most likely, for most situations, using 16 shades (4-bit PE) will be enough.

In real images, almost the entire area of the image is covered by a grid of cells that have functioned to change the gray code from the previous current frame. This is because in a real video, when digitizing, each pixel may not receive the same color codes, even if no changes at this point have occurred. For human vision, for example, the color RGB (180,180,180) will not differ from RGB (182,181,179), both of them will be light gray. But programmatically, these are two different colors with different codes.

In order for PE to not work, it is enough to enter the sensitivity threshold for such "identical" colors. All changes in the pixel color code that are less than this threshold should be counted in which no movement has been detected and vice versa for points whose color code changes more than the threshold level—count those that have recorded motion.

Using different sensitivity thresholds for the same visual scene, different results can be achieved depending on the problem statement. In this regard, certain rules are distinguished, using which it is possible to additionally build a special system for controlling the cellular environment, which itself automatically sets the desired sensitivity threshold.

One of these rules can be used as a percentage of the number of PEs that recorded the movement from the total number of all PEs of the medium to which the images were submitted (a moving object has a certain predetermined maximum size).

Also, the presence of rarely scattered PEs that have recorded movement along the entire image plane can be taken into account, because a moving object usually occupies a certain specific place. It should start with small thresholds of sensitivity and constantly increase it until a more or less localized cluster of moving points is obtained.

## 6 Experiment

For the experiment, the CASIA Gait base was used [11] (http://www.cbsr.ia.ac.cn/english/Gait%20Databases.asp). Software was developed that implemented the allocation of information pixels for motion detection, as well as software for the formation of a set of FAI for each frame of the video image. Initially, a new video

**Fig. 11** Formation of a video sequence with selected pixels

was formed for each set of files used. As a result, a set of BMP format files was generated, each of which displayed an image of selected pixels that detected movement (Fig. 11).

Each newly formed frame of the video sequence was binarized and determined by the FAI for different directions of shift of a copy. A database was created for each gait of person for each angle of the camcorder disposition. As already mentioned, the database was formed from codes representing a set of average FAI values for each generated video sequence. From the generated codes, codes were selected and compared with codes of other video sequences. Moreover, the beginning of the movement often does not coincide with the beginning of the movement in the generated code. Therefore, the program implemented a simultaneous shift of the codes of the reference database. In fact, the code of the input sequence was compared with each shifted code of each reference identifier from the database. The frame of the video sequence was determined, which is represented by one of the extreme average values for the compared codes, and the analysis of the obtained average value curve was carried out.

During the experiment, various confidence intervals were used that showed high identification results. Experimentally, the necessary values of the trust interval were determined that give the greatest accuracy of identification. A trust interval of ten pixels was determined for the database used. At this trust interval, an acceptable number of matches in the code was obtained when identifying "their own" (more than 69%) and a small number of matches (less than 10% matches) were obtained when identifying the "impostor."

## 7 Conclusion

Biometric identification of the person by the dynamics of his gait does not yet have accurate and common methods and means, which with high productivity and confidence, gave an accurate result. Video gait analysis requires a lot of effort to achieve success in solving this problem. There are a number of problems that do not allow achieving high results in this area at this stage. This chapter presents the video processing technology for biometric identification of a person by the dynamics of his gait. The proposed method consists of two main stages: the allocation of informational pixels of movement and the formation of the function of intersecting areas. The introduction of this technology has improved the characteristics of biometric identification, which are as follows:

1. There is no need to determine the location of the image of the human body in each video frame.

2. The method has no restrictions for large-scale changes.
3. The ability to accurately determine the repetition period of movements for any location of the camera.
4. The method is independent of the distance to the camcorder.
5. The method can be applied to any images of moving objects.
6. No large learning set for learning.
7. The change in gait speed is quickly detected.

Based on experimental studies, it was possible to determine trust intervals that give maximum identification efficiency. In the future, it is planned to conduct experiments for various locations of a person in front of video cameras and with various images of frames of a video sequence.

## References


1. Nambiar, A., Bernardino, A., & Nascimento, J. C. (2019). Gait-based person re-identification: A survey. *ACM Computing Surveys, 52*(2), 33.
2. Fairhurst, M. (2019). *Biometrics: A very short introduction (very short introductions)*. Oxford: Oxford University Press.
3. Das, R. (2018). *The science of biometrics: Security Technology for Identity Verification*. London: Routledge.
4. Nixon, M. S., Tan, T., & Chellappa, R. (2010). *Human identification based on gait (international series on biometrics book 4)*. New York: Springer.
5. Sokolova, A., & Konushin, A. (2018). Review of video gait recognition methods. *Computer Vision. GraphiCon, 2018*, 234–237.
6. Connor, P., & Ross, A. (2018). Biometric recognition by gait: A survey of modalities and features. *Computer Vision and Image Understanding, 167*, 1–27.
7. Figueiredo, J., Santos, C. P., & Moreno, J. C. (2018). Automatic recognition of gait patterns in human motor disorders using machine learning: A review. *Medical Engineering & Physics, 53*, 1–13.
8. Pradhan, C., Wuehr, M., Akrami, F., et al. (2015). Automated classification of neurological disorders of gait using spatio-temporal gait parameters. *Journal of Electromyography and Kinesiology, 25*(2), 413–422.
9. Blanca, N. (2017). Automatic learning of gait signatures for people identification. *Advances in Computational Intelligence, 2017*, 257–227.
10. Padole, C., & Proenc, H. (2017). An aperiodic feature representation for gait recognition in cross-view scenarios for unconstrained biometrics. Pattern anal Applic. *Pattern Analysis & Applications., 20*(1), 73–86.
11. Zheng, S., Zhang, J., Huang, K., He, R., & Tan, T. (2011). Robust view transformation model for gait recognition. In *Proceedings of the IEEE international conference on image processing*. Piscatawa, NJ: IEEE Operations Center.
12. Bilan, S., Al-zoubi, S. I., Yuzhakov, S., & Bilan, M. (2018). Description and recognition of symmetrical and freely oriented images based on parallel shift technology. In *5th international conference on mathematics and computers in sciences and industry (MCSI)* (pp. 86–91). Piscataway, NJ: IEEE.
13. Bilan, S., & Yuzhakov, S. (2018). *Image processing and pattern recognition based on parallel shift technology*. Boca Raton, FL: CRC Press/Taylor & Francis Group.
14. Yuzhakov, S., & Bilan, S. (2019). Identification system for moving objects based on parallel shift technology. In S. M. Bilan & S. I. Al-Zoubi (Eds.), *Handbook of research on intelligent data processing and information security systems* (pp. 289–329). Hershey, PA: IGI Global.

# User Identification Using Images of the Handwritten Characters Based on Cellular Automata and Radon Transform

**Stepan Bilan, Ruslan Motornyuk, Sergii Bilan, and Olga Galan**

## 1 Introduction

Handwriting recognition is currently given much attention by specialists. There are many known works in this area [1–8]. As a rule, well-known works describe the proposed methods and means for recognizing handwritten images. In additional, it is not determined to whom this text belongs. The author who wrote the text is not determined. High results in area of a text recognition are achieved. However, there are areas of human activity (banks, forensics, access systems, etc.) where the task is to biometricly identify the person who reproduces this text.

You can identify or authenticate a person using a static image of text fragments, as well as the dynamics of text playback over time. The static image of the handwritten text makes it possible to describe only the geometric shape of the curves forming the handwritten text. The slopes of the curves are taken into account and their various geometric quantities are measured and calculated. As a rule, well-known methods are applied to such images, which are based on the selection of characteristic features on the image, by the quantitative values of which they decide on a text symbol or a group of characters [9–11]. Each method uses a set of characteristic features with which it can function normally. Such methods are dependent on a set of characteristic features and on the shape of the images. Such methods

S. Bilan (✉) · O. Galan
State University of Infrastructure and Technology, Kyiv, Ukraine
e-mail: bilan_sm@gsuite.duit.edu.ua

R. Motornyuk
PU "Kiev Department" Branch of the Main Information and Computing Center of the JSC "Ukrzaliznytsya", Kyiv, Ukraine

S. Bilan
Onseo Company, Vinnytsia, Ukraine



S. Bilan et al. (eds.), *Biometric Identification Technologies Based on Modern Data Mining Methods*, https://doi.org/10.1007/978-3-030-48378-4_6

cannot be used for other classes of images and are not universal. The recognition system developer himself determines the characteristic features and develops a method aimed at their selection. Such a situation sharply limits the existing methods and means since in the world there are many languages and alphabets of different peoples. Therefore, the task is to create independent methods and means of recognizing handwritten text and identifying its author.

An analysis of the dynamics of handwriting playback identifies a person by the dynamics of the movement of the hand holding the writing element during the reproduction of handwritten text. These methods extract dynamic characteristic features during creating handwriting imaging. As a rule, such characteristic features include the coordinates of the direction of movement of the writing element at a certain point in time. Lengths of lines without tear-off movement of the writing element are also taken into account.

If for a person's identification method it is easy to fake a handwritten image, the dynamics of creating such a text is much more difficult to fake. Therefore, the method of identifying a person from static images of handwritten text in access systems to objects or information is not applied. However, in forensics this method is the most commonly used.

In this paper, we solve the problem of creating a method acceptable for handwritten characters of any alphabet and with a universal set of characteristic features that allows us to identify a person from statistical images of handwritten text. To solve this problem, the Radon transform implemented on a cellular automaton (CA) with hexagonal coating is used.

The problem is also solved by extracting common characteristic features for all types of handwritten image images by constructing six projections of the Radon transform and by further analyzing the quantitative characteristics of these projections for each image.

## 2 The Radon Transform Based on a Hexagonal Lattice

The method described in this chapter uses a cellular automaton with a hexagonal cell coating [12–14]. On the basis of such a CA the Radon transform is implemented [15, 16]. With hexagonal coating, each cell has uniquely six neighbors (Fig. 1), which forms six-connectedness.

The works [14, 17–20] show the advantages of using a hexagonal lattice. The main advantage is that when using the Radon transform, the number of projections

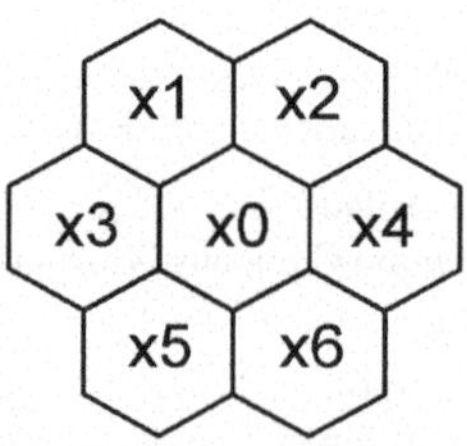


**Fig. 1** Coding neighbors on the hexagonal lattice

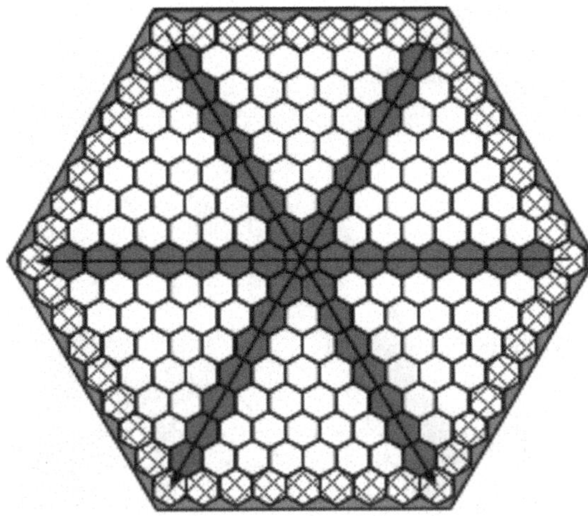

**Fig. 2** The main directions of the hexagonal lattice

is three in directions $0^0$, $60^0$, and $120^0$ [14]. These three projections represent the characteristic features of an image based on the Radon transform and give a complete description and restoration of images. The main directions of the hexagonal lattice in Fig. 2 are presented.

According to these directions, three projections are formed without gaps between cells (Fig. 3).

Orthogonal coating has only two directions for constructing projections of the Radon transform. However, these two projections are not enough to extract the characteristic features of the images, since they do not allow complete restoration of the original image. A sufficient minimum number of projections corresponds to three in directions $0^0$, $60^0$, and $120^0$. Figures 2 and 3 show the main directions for constructing the necessary three projections of the Radon transform. In this case, there is the possibility of increasing the number of projections of the Radon transform by using the symmetry of rotation of the regular hexagon [12–14]. In this case, the number of projections increases to six.

For this, the concepts of additional directions of the hexagonal lattice are introduced [13, 14]. These directions intersect in the central cell of the hexagonal lattice. They pass through the vertices of the central cell of the cellular automaton and are conditionally indicated by the transition from the boundary between one pair of neighboring cells to the border between another opposite pair of neighboring cells. Such additional directions for constructing projections of the Radon transform are angles $30^0$, $90^0$, and $150^0$ (Fig. 4).

The combination of the main and additional directions gives six directions ($0^0$, $30^0$, $60^0$, $90^0$, $120^0$, and $150^0$) for constructing projections of the Radon transform on a hexagonal lattice (Fig. 5). From the direction $180^0$ to $360^0$, the directions are symmetrical to the directions described above.

The main and additional directions on the hexagonal lattice allow to describe any image and reproduce it effectively.

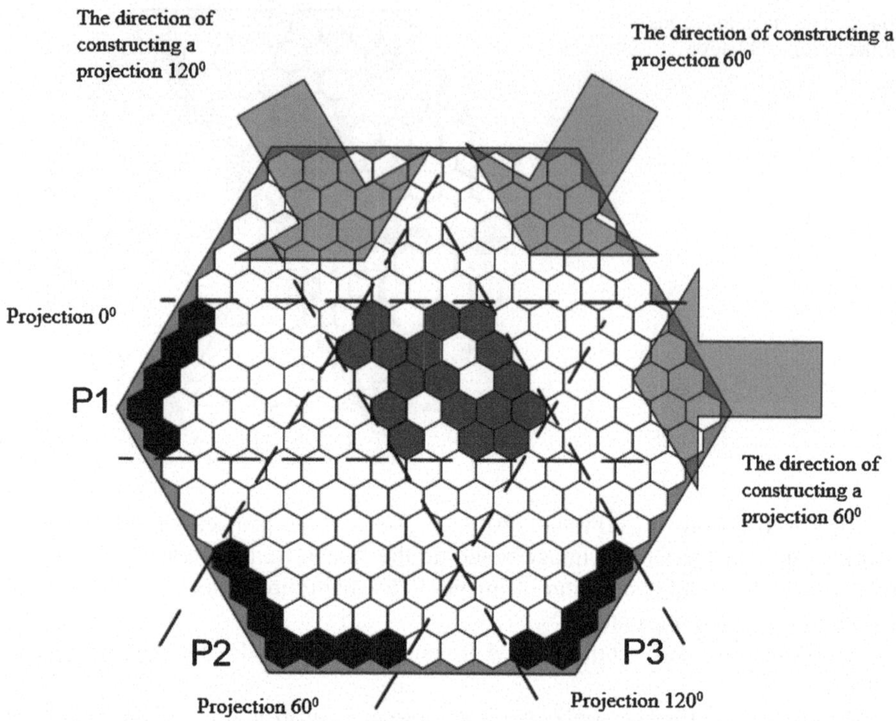


**Fig. 3** An example of the formation of projections on a hexagonal lattice

**Fig. 4** Additional directions of the hexagonal lattice that correspond to the corners $30^0$, $90^0$, and $150^0$

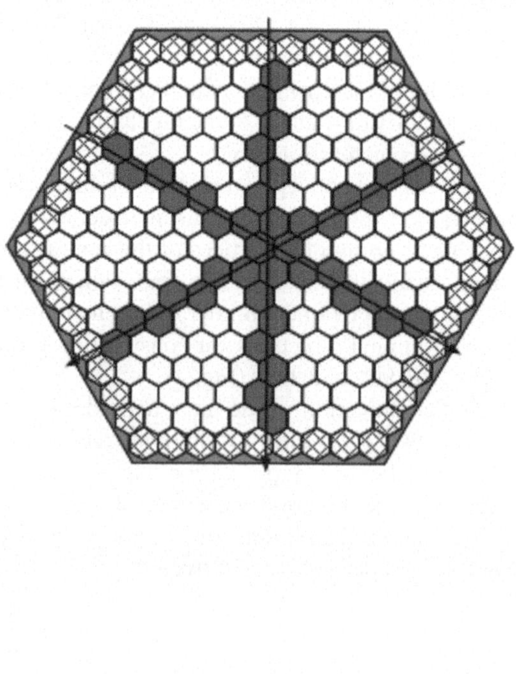

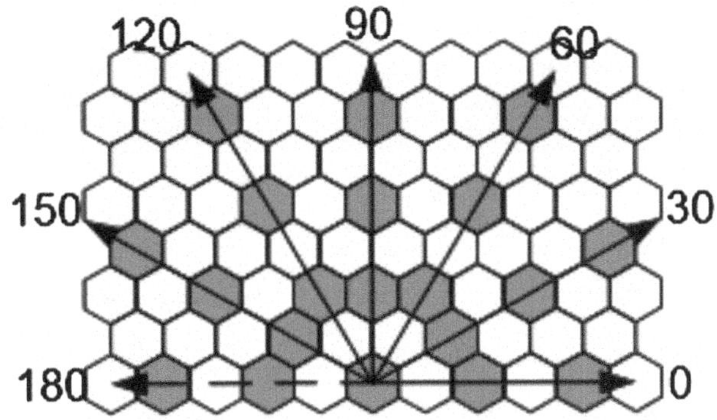


**Fig. 5** The main and additional directions for constructing projections of the Radon transform

## 3 Description of Handwriting Image Recognition Method

The image is described using characteristic features, which are projections of the Radon transform in the corresponding directions. Radon projection is determined by the sum of the cells belonging to the image and lying on the same lines of the projection construction direction. Each cell belonging to the image takes part in the formation of the sum in each of the six directions.

Addition begins with the extreme cell belonging to the direction line. On each line of direction, the brightness values of cells lying in one straight direction are summed. Each projection of the Radon transform consists of an ordered set of sums of brightness values of the corresponding cells belonging to the image. Obtaining such projections can be achieved by shifting the image in six directions to the edges of the lattice and counting the image cells in the corresponding position of the hexagonal lattice. The areas for constructing projections of the Radon transform in the main and additional directions in Fig. 6 are presented.

Given this arrangement, the number of accumulating adders located at the edges of the hexagonal lattice was determined. The general scheme of accumulating adders place in Fig. 7 is shown.

According to the developed model, the image is fully described by six projections of the Radon transform. From such projections a database is formed. Each set of six projections of the Radon transform is assigned a corresponding identifier. This happens when the image at the input of the identification system is known, i.e. initial system learning. Projections with an identifier are reference. The more reference projections with one identifier in the database, the more accurate the identification.

After the formation of the reference database, the system is ready for identification. Handwritten images are input. Separate symbols are selected, as well as their combinations, and six projections of the Radon transform are formed. The generated projections of the Radon transform are compared with the reference projections in the database. If the result of the comparison corresponds to the rules of accurate identification, for the selected reference projections, then the person is

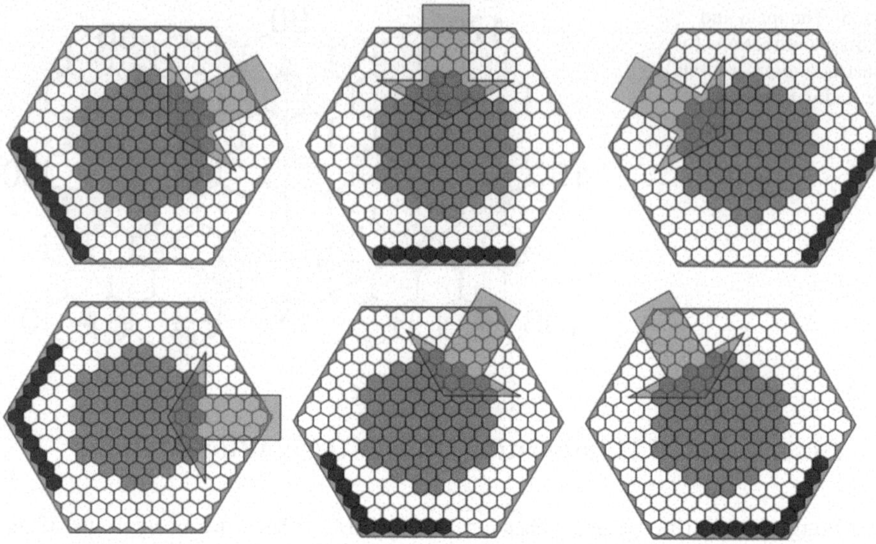

**Fig. 6** Areas for constructing projections of the Radon transform for the main and additional directions

**Fig. 7** The schema of the emplacement of the accumulating adders

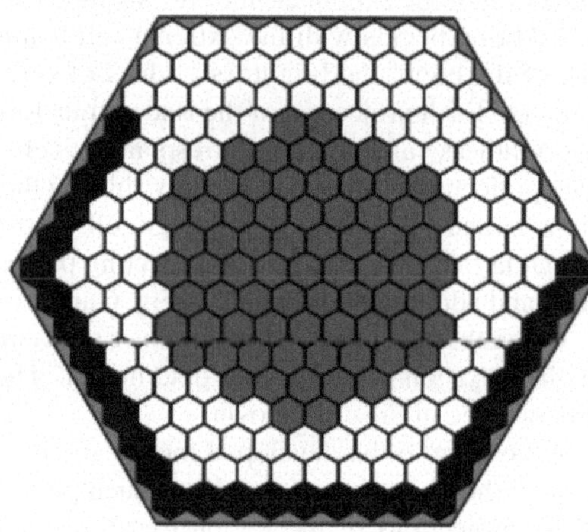

identified. If identification is carried out with 100% certainty, then the generated projection is added to the database to the reference projections with the corresponding identifier. Thus, the system extends the database in the identification process.

However, in the process of forming projections of the Radon transform, threshold processing of the handwritten image is initially performed, as a result of which a binary image is formed. This image has black pixels, which are separate elements and can affect projection shapes. Such projections introduce significant deviations from the quantitative values of the reference projections of the Radon transform. An

**Fig. 8** An example of the removal of extra pixels that do not belong to the image of the character

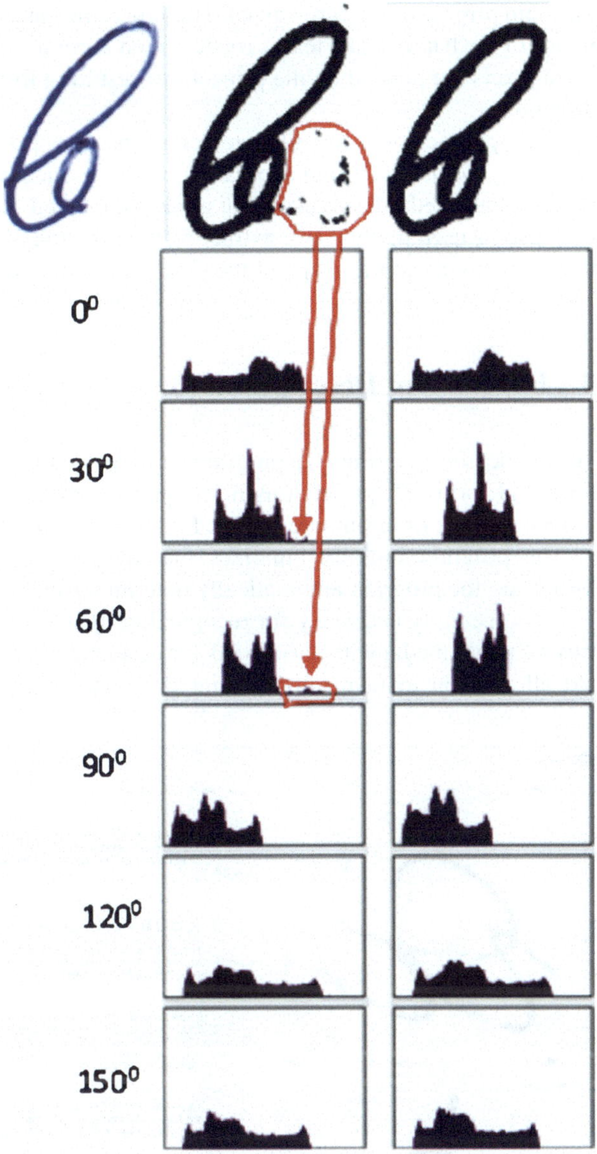


example of the initial symbol image and symbol image without extra black pixels and their projection of the Radon transform in Fig. 8 is presented.

In this example, the extra pixels are highlighted in red, obtained after binarization, as well as the projection of the Radon transform corresponding to the extra pixels in the image. Such projection sections are clearly selected on the projections obtained in the directions $30^0$ and $60^0$. They are insignificant and are separated from the main projection by zero values. Therefore, they are defined as superfluous. If the image of the handwritten text is rotated, then sections of the projections defining the extra pixels are present on the projections resulting from the formation in the other direction.

This approach does not allow to determine extra pixels located inside the image, since they do not stand out separately on the formed projections. To identify such pixels, other methods are used that are implemented on cellular automata [21, 22]. Can also be used methods for extracting the skeleton of the image, which allow you to clearly highlight the shape of the character in the image [23].

## 4 Experiment Description

To conduct the experiment, a program was created that implements the Radon transform and forms six projections in graphical form and in the form of quantitative sequences. The program interface in Fig. 9 is shown.

The program initially binarizes the image at a given brightness threshold. Moreover, the program automatically determines the required brightness threshold. This is especially necessary for recognizing images of handwritten text printed on various paper media with various inks. Automatic selection of the brightness threshold allows you to take into account the force of pressing the writing element.

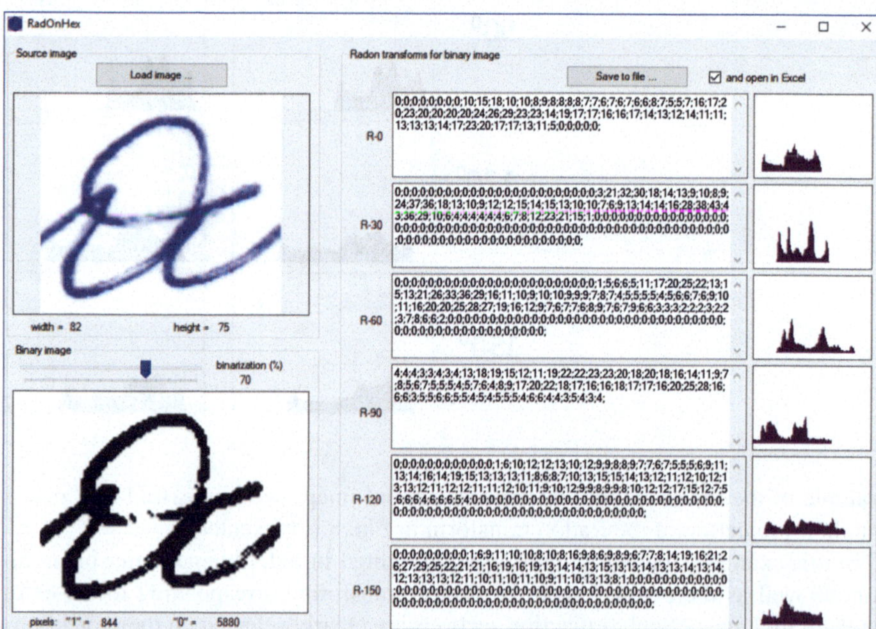


Fig. 9 The program interface that forms the projection of the Radon transform

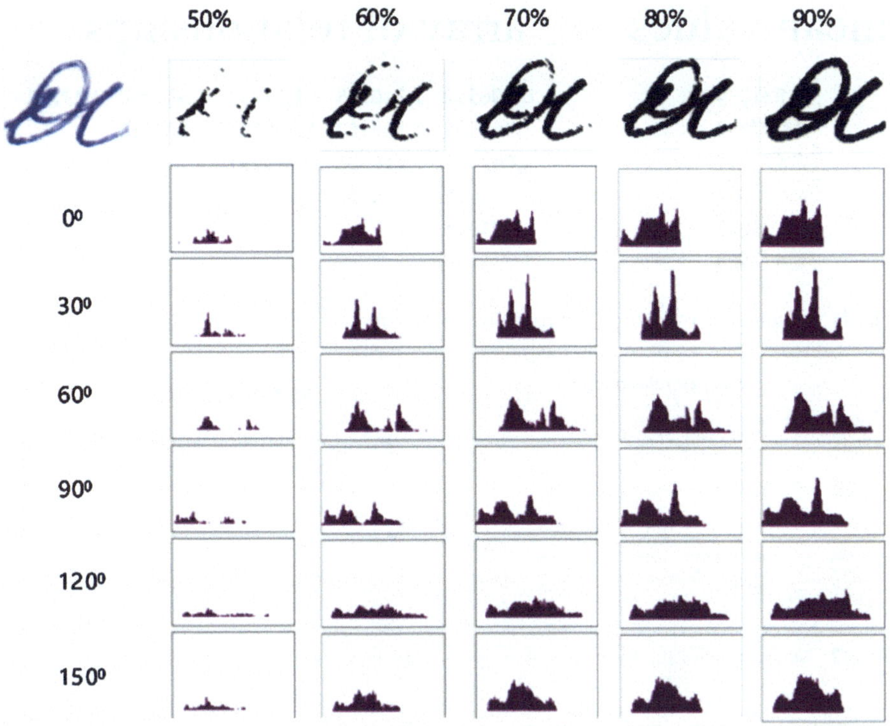


**Fig. 10** Projections of the Radon transform for the image of the symbol "a" at thresholds of brightness of 50%, 60%, 70%, 80%, and 90%

Figure 10 shows projections of the Radon transform of the same image for different brightness thresholds.

Different brightness thresholds give different forms of image. For lower brightness thresholds, there are gaps in the outlines of characters, where clicking on the writing element was weaker. For a higher luminance threshold (for example, 90%), the symbol image is displayed without gaps. In binary images with a large brightness threshold, only the shape of the reproduced handwritten image is evaluated.

During the experiment, a database was generated in which the projection codes of the Radon transform of the handwritten images of each person participating in the experiment were recorded. Moreover, the projection codes of the Radon transform were divided into groups as follows. The database was divided into sets of codes. Each set was assigned the identifier of the person who reproduced the corresponding handwritten text.

Each set identifying a person was divided into subsets of codes. Each subset contained projection codes for the Radon transform for images of combinations of handwritten characters and for individual characters.

To increase the accuracy of identification, at least ten reference projections of Radon transforms were formed for each fragment of the handwritten text. Moreover,

**mean values**

| |
|---|
| 9,931034 |
| 11,07692 |
| 7,578947 |
| 7,2 |
| 6,26087 |
| 8,470588 |

**array of relationships**

| | | | | |
|---|---|---|---|---|
| 0,896552 | 1,310345 | 1,37931 | 1,586207 | 1,172414 |
| 1,461538 | 1,538462 | 1,769231 | 1,307692 | |
| 1,052632 | 1,210526 | 0,894737 | | |
| 1,15 | 0,85 | | | |
| 0,73913 | | | | |

**Fig. 11** Example of a vector of average values and an array of relations

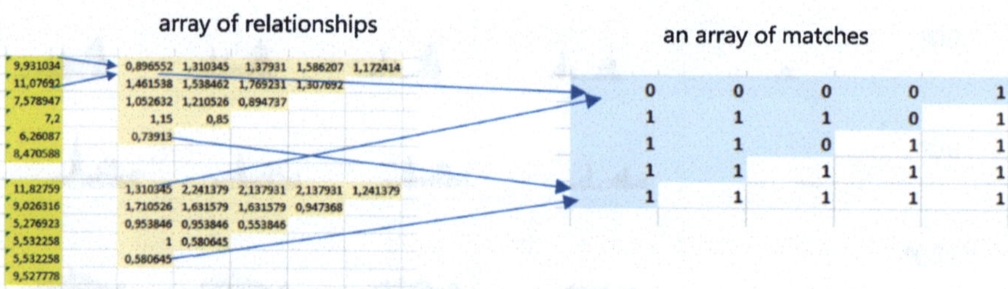


**Fig. 12** An example of the formation of arrays of coincidences

fragments of handwritten text were selected from different parts of the text (beginning, middle, end of text).

The average values for each projection of the Radon transform were determined from the images of each fragment of the handwritten text. The average values were selected from the array of projection numbers of the Radon transform within the boundaries of the beginning and end of each projection. The remaining zeros in each projection were not taken into account when calculating the average value. Thus, for each image a vector of average values was formed. An example of such a set of relationships on Fig. 11 is presented.

Comparison of arrays of relations for the images of each fragment of handwritten text was made and arrays of coincidences were formed. Match arrays consist of values 0 and 1. To form arrays of coincidences from each number of one array of relations, the corresponding numbers were subtracted from another array of relations. If the value of the subtraction result exceeded the value of the confidence interval from the corresponding position of the coincidence array, 0 was formed, and if the result of the subtraction did not exceed the value of the confidence interval, then 1. An example of the formation of coincidence arrays on Fig. 12 is presented.

For ten relationship arrays, 45 coincidence arrays were formed. All 1 were counted in the corresponding positions of coincidence arrays for each position.

Based on the generated database, the identification was carried out as follows.

Images of handwritten fragments were fed to the input of an identification system. Their binarization was carried out, extra pixels in the binary image were removed and projections of the Radon transform were formed. The obtained projections were compared with the reference ones and the percentage of coincidences was determined. If the average value of the percentage of coincidences exceeded the selected threshold for one of the many reference values, then a specific identifier identified the user. If the average value of the percentage of coincidences did not exceed the specified threshold, then the user was identified as an impostor.

The experiment was conducted for confidence intervals of 0.1; 0.2; 0.3, and 0.4. The most optimal confidence intervals are 0.2 and 0.3. At these confidence intervals, the difference between the average values for the identification of "their own" and "impostor" is 20%. At the same time, the average value in the identification of "one's own" is always more than 55%, and in determining the "impostor" the difference is less than 50%.

Such an insignificant difference is characterized by the fact that for different handwritings some projections of the Radon transform can be close in quantitative values. Such an insignificant difference is characterized by the fact that for different handwritings some projections of the Radon transform can be close in quantitative values.

Also in the experiment, images of various fragments of the handwritten test were used. Separate characters and their combinations of two, three, etc. were selected. For them, the projections of the Radon transform were determined and the percentage of coincidences was calculated. The experiment confirmed the proximity of the average values of coincidences for images of various combinations of pulses. Figure 13 shows the average value of the percentage of matches for different combinations of handwritten symbol groups of one person, as well as the average value of the same combinations of symbol groups for one impostor.

The blue values are used to identify "their," and the brown are the values used to identify the "impostor." The graph is presented for arbitrary groups of characters from 1 to 5. Combinations of three characters show the greatest difference. The difference is 40%. The experiment showed that no significant discrepancies were observed, which confirms the effectiveness of the proposed method.

## 5 Conclusion

The paper analyzes existing methods for recognizing handwritten images, as well as methods for identifying a person. Based on the analysis, a personality identification method using the Radon transform and its implementation on a cellular automaton with hexagonal coating was proposed. This made it possible to introduce additional projections of the Radon transform, which significantly increased the accuracy of identification. Due to the proposed method for processing the numerical values of the projections of the Radon transform, the optimal confidence interval is determined, which gives an accurate identification of a person from handwritten images. As a result of the experiment, the value of the percentage of coincidences was determined, which distinguishes "ours" from "impostor." Combinations of groups of 3 and 4 characters each showed the best result.

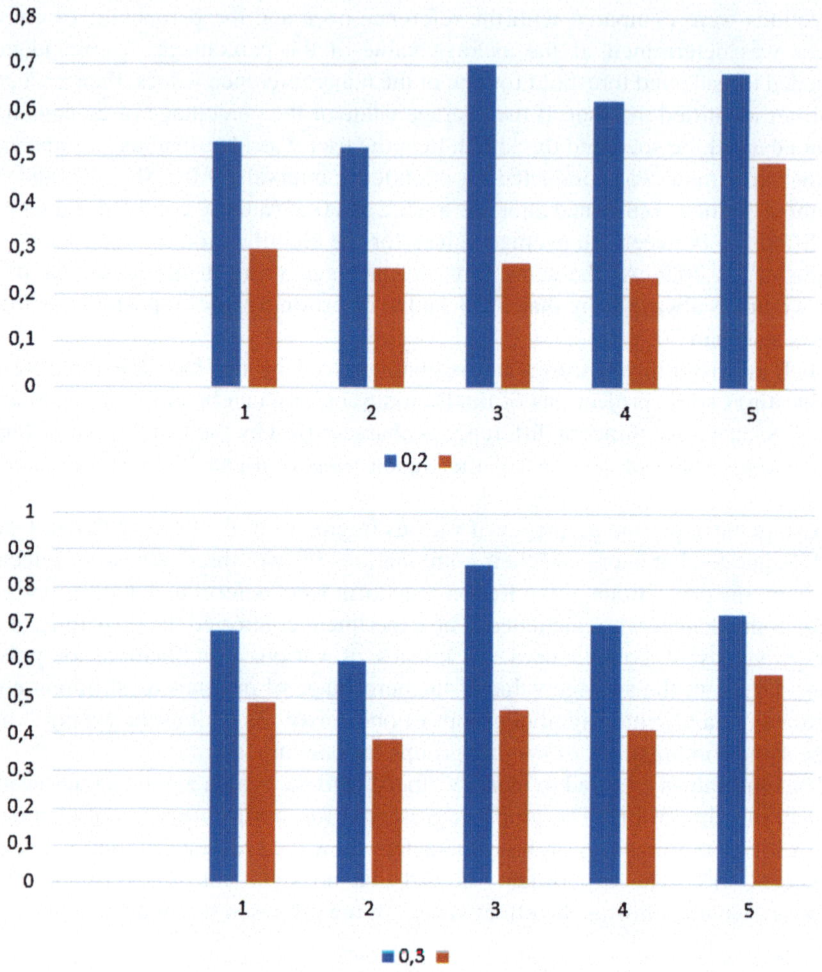


**Fig. 13** The dependence of the average values of coincidences for various combinations of groups of characters for "their" and "impostor" and the binarization threshold of 70%

In the future, the authors plan to conduct research in the field of biometric identification on the dynamics of writing a text by a person. For this, it is planned to use the same technique based on the Radon transform and the hexagonal lattice.

## References

1. Lorigo, L. M., & Govindaraju, V. (2006). Offline Arabic handwriting recognition: A survey. *IEEE Transactions on Pattern Analysis and Machine Intelligence (TPAMI), 28*(5), 712–724.
2. Vinciarelli, A. (2002). A survey on off-line cursive word recognition. *Pattern Recognition, 35*(7), 1433–1446.

3. Graves, A., Fernandez, S., Gomez, F., & Schmidhuber, J. (2006). Connectionist temporal classification: Labelling unsegmented sequence data with recurrent neural networks. In *23rd International Conference on Machine Learning* (pp. 369–376). New York: ACM.
4. Graves, A., Liwicki, M., Fernandez, S., Bertolami, R., Bunke, H., & Schmidhuber, J. (2009). A novel connectionist system for unconstrained handwriting recognition. *IEEE Transactions on Pattern Analysis and Machine Intelligence (TPAMI), 31*(5), 855–868.
5. Zamora-Martinez, F., Frinken, V., Espãna-Boquera, S., Castro-Bleda, M. J., Fischer, A., & Bunke, H. (2014). Neural network language models for off-line handwriting recognition. *Pattern Recognition, 47*(4), 1642–1652.
6. Wigington, C., Stewart, S., Davis, B., Barrett, W., Price, B. adn Cohen, S. (2017). Data augmentation for recognition of handwritten words and lines using a CNN-LSTM network. 14th International Conference on Document Analysis and Recognition (ICDAR): 639–645.
7. Snchez, J. A., Romero, V., Toselli, A. H., & Vidal, E. (2016). ICFHR2016 competition on handwritten text recognition on the READ dataset. In *15th International Conference on Frontiers in Handwriting recognition (ICFHR)* (pp. 630–635). Piscataway, NJ: IEEE. https://doi.org/10.1109/ICFHR.2016.0120.
8. Puigcerver, J. (2017). Are multidimensional recurrent layers really necessary for handwritten text recognition? In *14th International Conference on Document Analysis and Recognition (ICDAR)* (pp. 67, IEEE–72). Piscataway, NJ. https://doi.org/10.1109/ICDAR.2017.20.
9. Rane, J. (2018). *Person identification system using handwritten signatures by ANN: a neural network perspective*. LAP LAMBERT Academic Publishing.
10. Dabra, S., & Agrawal, S. (2013). *Identification of handwritten characters using machine learning*. LAP LAMBERT Academic Publishing.
11. Li, H., Doermann, D., & Zheng, Y. (2008). *Handwritten document image processing: Identification, matching, and indexing of handwriting in Noisy document images*. Saarbrücken: VDM Verlag.
12. Bilan, S., Motornyuk, R., & Bilan, S. (2014). Method of hardware selection of characteristic features based on radon transformation and not sensitive to rotation, shifting and scale of the input images. *Advances in Image and Video Processing, 2*(4), 12–23.
13. Belan, S. N., & Motornyuk, R. L. (2013). Extraction of characteristic features of images with the help of the radon transform and its hardware implementation in terms of cellular automata. *Cybernetics and Systems Analysis, 49*(1), 7–14.
14. Motornyuk, R.L. (2013). Computer-aided methods for identifying images of moving objects based on cellular automata with a hexagonal coating. Dissertation for the degree of candidate of technical sciences (UDC 004.932: 519.713 (043.3)), Kiev: SUIT, 2013.
15. Rubin, B. (2015). *Introduction to radon transforms: With elements of fractional Calculus and harmonic analysis (encyclopedia of mathematics and its applications)* (1st ed.). Cambridge: Cambridge University Press.
16. Helgason, S. (2014). *Integral geometry and radon transforms*. Berlin: Springer.
17. Dudgeon, D. E., & Mersereau, R. M. (1984). *Multidimensional Digital Signal Processing*. Englewood Cliffs, NJ: Printice-Hall.
18. Mersereau, R. M. (1979). The processing of hexagonally sampled two-dimensional signals, proc. *IEEE, 67*(6), 930–949.
19. Staunton, R. C. (1989). Hexagonal image sampling, a practical proposition. Proc. *SPIE, 1008*, 23–27.
20. Nicoladie, D. T. (2014). Hexagonal pixel-array for efficient spatial computation for motion-detection pre-processing of visual scenes. *Advances in Image and Video Processing, 2*(2), 26–36.
21. Wolfram, S. (2002). *A new kind of science*. Champaign, IL: Wolfram Media.
22. Bilan, S. (2017). *Formation methods, models, and hardware implementation of pseudorandom number generators: Emerging research and opportunities*. Hershey, PA: IGI Global.
23. Motornyuk, R. L., & Bilan, S. (2019). Methods for Extracting the Skeleton of an Image Based on Cellular Automata With a Hexagonal Coating Form and Radon Transform. In S. M. Bilan & S. I. Al-Zoubi (Eds.), *Handbook of Research on Intelligent Data Processing and Information Security Systems* (pp. 289–329). Hershey, PA: IGI Global.

# Automated Calculation of Fundamental Matrix from Stereo Images from a Different Point of View

**Ahmed Chater and Abdelali Lasfar**

## 1 Introduction

The epipolar geometry of a scene illustrates the connection between different images of the same scene taken from various vantage points by producing the intrinsic projective geometry between the views [1]. The calculus of the fundamental matrix (F) that establishes the epipolar geometry is based on the calibration of the camera [2] and self-calibration [2, 3], 3D projective reconstruction [2, 4], motion analysis object matching and tracking [2], 3D target localization, and 3D tracking of personnel [5]. The calculation of F requires at least seven points of correspondence. Several manual methods of selecting points have been proposed, but they produce many errors, so it is not practical for them to process a string of images or video frames. Several automatic point detection techniques, such as Harris corner detector [6], (FAST) [7], Scale Invariant Robust Features (SIFT) detector [8], and Speeded Up Robust Features (SURF) detector [9], have been used to detect significant points in different stereo images. These images are then automatically matched in opposition views by applying point-matching algorithms. SIFT and SURF detectors, which combine a descriptor vector for each significant point, behave better at this level with incorrect ratio parameters and correspondences. Moreover, it has been shown that SURF is faster than SIFT in terms of calculating F, but identify fewer significant points to SIFT in [10]. After determining the correspondences between the points using the different detectors, the fundamental matrix is calculated using the standard eight-point method [11], SURF gave the points a solution to calculate F. This algorithm is sensitive to the determination of the matrix F [12, 13]. Robust statistical tools such as least median of squares (LMedS) [14] and RANdom SAmple Consensus (RANSAC) [15] are often employed in this level to eliminate outliers

A. Chater · A. Lasfar (✉)
Laboratory of System Analysis, Information Processing and Industry Management,
High School of Technology SALE Mohammed V University in Rabat, Rabat, Moroccot



S. Bilan et al. (eds.), *Biometric Identification Technologies Based on Modern Data Mining Methods*, https://doi.org/10.1007/978-3-030-48378-4_7

[16]. In this work, we present a method that increases the speed of calculating F and combines the SURF detector and the normalized eight-point algorithm. After an application of this method with a modification of the threshold on three pairs of images were been conducted with a different change (rotation, illumination, and displacement of objects). The simulations give good results in terms of F calculation speed compared to the other methods (Harris, SIFT, FAST + RANSAC algorithm + eight points standardized) [17].

## 2 Correspondence Matching

All methods of estimating the fundamental matrix require a number of points matching as an input element [18]. Several detectors and descriptors have been reported in the literature, and in this article we studied the following four: Harris, SIFT [10], FAST [7], and SURF [9, 19], and binary descriptors [20, 21]. Of these techniques, the detection of significant points, we chose the descriptors because the points of interest do not include enough data for a low-contrast match. According to Mikolajczyk Schmid [22], the comparison between the different detectors (SIFT, FAST, Harris and others) showed that SIFT gave good results in terms of variation of rotation and scale. Bauer et al. [11] the SIFT descriptor which allows detecting the number of descriptor higher than SURF [21]. Bay and al [11] have developed a detector (SURF) which increases the computation time compared to (SIFT).

## 3 Epipolar Geometry and the Calculation of the Fundamental Matrix

### *3.1 Epipolar Geometry*

Epipolar geometry is intrinsic to any two camera systems regardless of the model used for these cameras. It was introduced by Longuet-Higgins [22]. This geometry describes a geometric relationship between two images taken from two different angles of view. The precision of estimating epipolar geometry is very important because it determines the precision of the pairing algorithms between the points of a pair of stereoscopic images, and these algorithms often rely on prior knowledge of this geometry.

In Fig. 1 $M_i$ is a 3D point $O$ and $O'$, are camera centers, are epipoles, and $l$ and $l'$ e'l' are epipolars lines are O and O 'are the optical centers corresponding to the left and right cameras right respectively. Similarly, $m_i$ and $m_i'$, the two left and right projections, correspond to the same 3D point $M$ of the scene. The projection of $O'$ in the right image plane will be called the epipole ($e$). In the same way the projection of $o$ in the left image plane called epipole ($e'$) the coplanarity of the points $\left(O, O', m_i, m_i'\right)$ formulates the fundamental matrix from the right and left images. This constraint is called epipolar constraint.

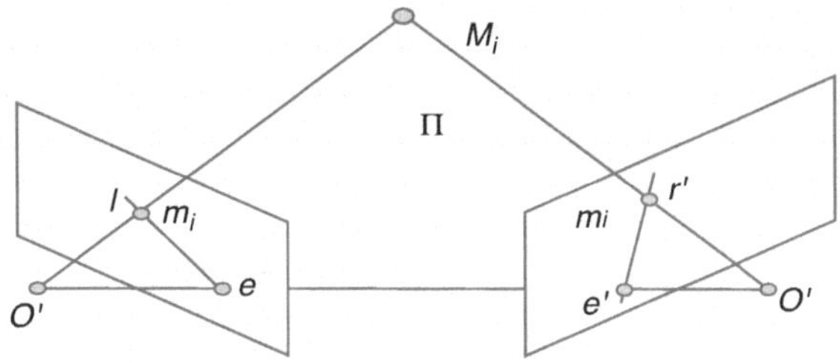


**Fig. 1** Epipolar geometry

### 3.2 *Fundamental Matrix*

The fundamental matrix gives the transformation by drawing a selected point in one of the images as an epipolar line on the other image, thus projecting a point on a line. Mathematically, the epipolar constraint is probabilistically translated by the fundamental matrix as indicated in the following equation:

$$m_i^{'T} F m_i = 0 \tag{1}$$

where F is a matrix of dimension 3 × 3 and of rank 2 and determined from F and zero, the Eq. (1) is the relationship which relies the points of the left image noted $m_i = (x_i, y_i, w_i)$ and points of the right image noted $m'_i = (x'_i, y'_i, w'_i)$.

## 4 Methods to Calculate the Fundamental Matrix

Many methods of estimating the fundamental matrix exist and most use only image point matches as data. These methods can be classified into linear methods, iterative, or robust [14]. Studies have shown that linear methods, and in particular the one introduced by Longuet-Higgins [22], are very sensitive to noise due to false matches. Iterative (or non linear) Methods allow outliers to be integrated and high-potential points to be cancelled by iteration. this technique is noise-resistant when images are noisy in [23], are able to give an accurate result with noisy images and can also take into account outliers. The following sections provide a summary of the principles used in each of these categories.

### 4.1 *Linear Method*

Equation (1) is the relation linking a point $m_i = (x_i, y_i, w_i)$ from the right image to a point $m'_i = (x'_i, y'_i, w'_i)$ from the left image. This equation can be rewritten as homogeneous linear coordinates in the following form:

$$Af = 0 \tag{2}$$

with $f = [f_{11}, f_{12}, f_{13}, f_{21}, f_{22}, f_{23}, f_{31}, f_{32}, f_{33}]^T$

$$A = \begin{pmatrix} x'_1 x_1 & x'_2 x_2 & . & x'_n x_n \\ x'_1 y_1 & x'_2 y_2 & . & x'_n y_n \\ x'_1 w_1 & x'_2 w_2 & . & x'_n w_n \\ y'_1 x_1 & y'_2 x_2 & . & y'_n x_n \\ y'_1 y_1 & y'_2 y_2 & . & y'_n y_n \\ y'_1 w_1 & y'_2 w_2 & . & y'_n w_n \\ y'_1 x_1 & y'_2 x_2 & . & y'_n x_n \\ w'_1 y_1 & w'_2 y_2 & . & w'_n y_n \\ w'_1 w_1 & w'_2 w_2 & . & w'_n w_n \end{pmatrix}^T$$

This set of linear equations makes it possible to establish the epipolar geometry in a given pair of images. The main advantage of this technique is it allows one to choose the best seven points for the estimation of F. However, this becomes a disadvantage when some points are badly located [24, 25]. In practice, there are more than 7 corresponding points. If we ignore the constraint on the rank of F, which is equal to 2, we can use the least-squares method to solve the following equation:

$$\min \sum_i \left( m_i'^T F m_i \right)^2 \tag{3}$$

By imposing a constraint making the norm of F equal to 1, the problem becomes a classic minimization problem [26] in the following form:

$$\begin{cases} \min_F \left\| A_n . F \right\|^2 \\ \left\| F \right\| = 1 \end{cases} \tag{4}$$

The resolution is then performed using the Barn multiplier technique as shown in Eq. (5):

$$\min_F \psi (F, \lambda) \tag{5}$$

The solution for F is the eigenvector that corresponds to the smallest eigenvalue λ. The estimation of F can be done more simply by using the eight-point algorithm. But the solution obtained is not necessarily optimal. On the other hand, the fundamental matrix of dimension (3 × 3) has two necessary conditions: its rank is 2 and determines equal zero. By using these characteristics related to the nature of the matrix, it is possible to improve the methods for estimating F. There is an a posteriori solution to find a matrix of null determinant from near F. The proximity of the two matrices is estimated by the so-called Frobenius norm [15]. To obtain the matrix, F is decomposed into the following form using a of singular value decomposition (SVD)-type technique: or S = diag $(\sigma_1, \sigma_2, \sigma_3)$ is a diagonal matrix with $\sigma_1 > \sigma_2 > \sigma_3$ and U and V are orthogonal matrices one can then demonstrate that the matrix $\hat{F} = U.\hat{S}.V^T$ with $\hat{S} = diag(\sigma_1, \sigma_2, 0)$ is a rank 2 matrix that minimizes the Fresenius norm of $F - \hat{F}$. This algorithm was perfected by Hartley [27] to make it even more robust. Thus, he proposed an algorithm: eight standardized points [14, 28, 29]. It has been shown that the application of the eight-point algorithm is often unstable. The solution proposed is to replace the origin in each image by the centroid of the paired points. Then a scaling factor is applied so that the mean norm of the vectors associated with the points is equal to the root of two. These two operations amount to multiplying the points of the left (right) image by a matrix (3 × 3). This approach has greatly improved the outcome of the eight-point method.

## 4.2 *Nonlinear Method*

This technique is based on measuring the distance between detected points and their epipolar line. See the preceding Eq. (1). It directly relates to the distance *d* from point $m_i$ to its epipolar right. A first idea is then to use a nonlinear criterion minimizing the sum [13, 30, 31], as shown in Eq. (6):

$$\sum_{i=1}^{n} d^2 (m_i, L_r) \tag{6}$$

with $d(m_i, L_r) = \frac{\left| m_i^{'T} F m_i \right|}{\sqrt{(F.m_i^{'})_1^2 + (F.m_i)_2^2}}$, where the term $(F.m_i^{'})_i$ is the *i*th element vector F which gives a symmetrical role to the two images and opener a coherent epipolar geometry and should minimize both creatures the distance between points and the epipolar line. This technique gives very good results compared to those obtained with linear methods. Although iterative methods are more accurate than linear and nonlinear methods, they cannot eliminate outliers.

## 4.3 *Iterative Method*

Iterative methods are called such because they are more tolerant to noisy data. RANSAC [15, 32], LMedS [14], and M-estimators [30] are three widely used robust methods reported in the literature. The first method calculates for each value of F the number of points that may be suitable (inliers) [25, 29, 33]. The matrix F chosen is that which maximizes this number. Once the aberrant points are eliminated, the matrix F is recalculated to obtain a better estimation. The LMeds method calculates for each estimation of F the Euclidean distance between the points and the epipolar lines, and the choice of F corresponds to the minimization of this distance. Although

M-estimator was inspired by the two preceding methods, it consists in dividing the detected points into two sets: inliers and quasi-inliers. The latter method is based on solving the following expression:

$$\min_F \sum_i w_i r_i^2 \tag{7}$$

$w_i$: weighting function

$$r_i = (x_i, y_i, w) \begin{bmatrix} f_{12} & f_{12} & f_{13} \\ f_{21} & f_{22} & f_{23} \\ f_{31} & f_{32} & f_{33} \end{bmatrix} (x_i', y_i', w')^T$$

$$r_i = f_{12}x_i x_i' + f_{21}y_i x_i' + f_{31}wx_i' + f_{21}x_i y_i' + f_{22}y_i y_i' + f_{23}wy_i' + f_{13}x_i w' + f_{23}y_i w' + f_{33}ww'$$

Huber [13, 25] has proposed the following expression for$w_i$:

$$w_i = w_i\left(p_i', p_i''\right) = \begin{cases} 1 & r \le \phi_i \sigma \\ \theta & \phi_i \sigma < r < \sigma \\ \dfrac{\theta\sigma}{|r_i|} & \sigma \le r_i < \varphi\sigma \\ 0 & \varphi\sigma < r \end{cases} \tag{8}$$

$\varnothing$ : Factor to ensure the boundary of the inliers and quasi-inliers.

The robust standard deviation $\sigma$ can be expressed as $\sigma = \dfrac{median(r_i)}{\lambda}$.

The following equation allows us to calculate the scale of error.

$\theta$: proportional factor whose scope is (0, 1);

$$E_{\text{sampson}} = \sum_i^n w_i \frac{\left(m_i^T F m_i'\right)^2}{\left(Fm_i\right)_1^2 + \left(Fm_i\right)_2^2 + \left(F^T m_i'\right)_1^2 + \left(F^T m_i'\right)_2^2} \tag{9}$$

where.

$(Fm_i)^2_j$, J = 1, 2 the square of the $j$th entry for the vector $(Fm_i)$, according to Liqiang Wang and [12].

Experiments have shown that the LMeds technique gives a better result than the RANSAC method in terms of accuracy [2]. LMeds and RANSAC are considered similar; they consist in selecting randomly a set of points to use for the approximation of the fundamental matrix. The difference between these two methods lies in the way they determine the chosen F. LMeds calculates F from the distance between points and the epipolar lines where it seeks to minimize the median. RANSAC calculates the matrix F from the number of inliers [33–35]. However, M-Estimator leads to a good result in the presence of Gaussian noise at the selected points of an image, and the robustness of this method is manifested in the reduction of aberrant values.

## 5 Simulation Results and Discussion

In this section, we apply the different points of interest detectors namely: Harris, FAST, SIFT and SURF (Fig. 2), then we standardize the exit points and combine them with the RANSAC algorithm which allows to classify the points in (inliers and other) in order to calculate the fundamental matrix with different variations of the scene. Then the fundamental matrix is computed without and with modification of the uniqueness threshold.

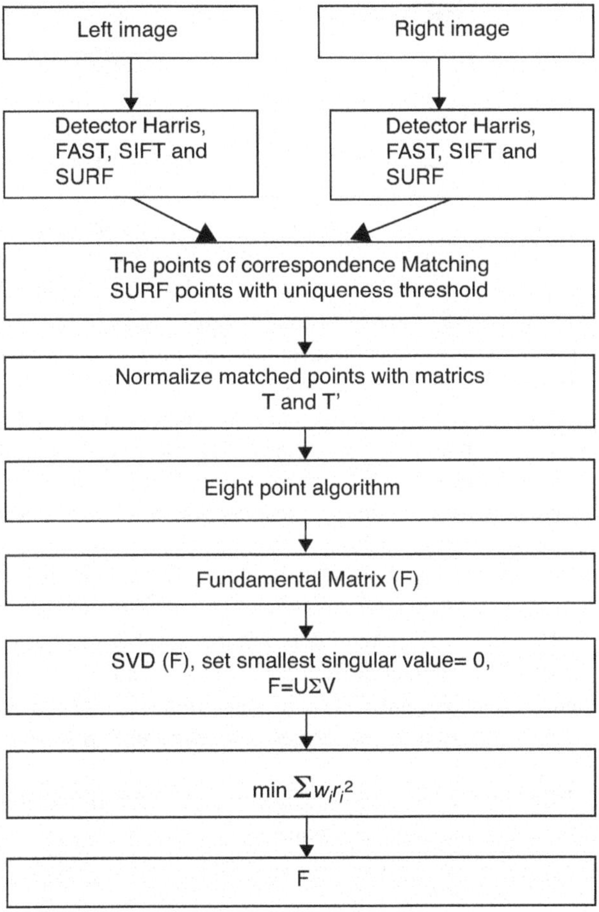


**Fig. 2** Methodology

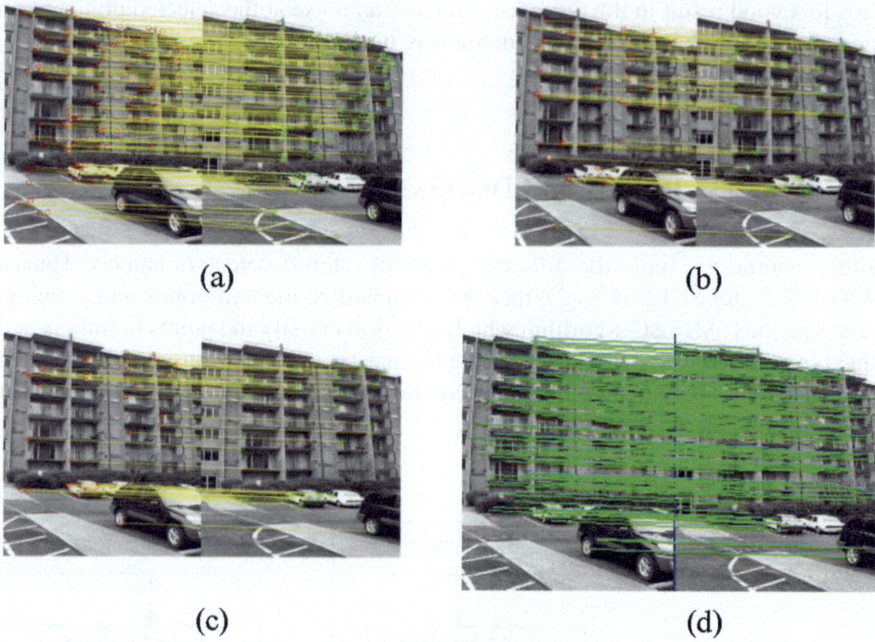

**Fig. 3** Correspondence of original image in motion: SURF (**a**), Harris (**b**), FAST (**c**), and SIFT (**d**)

Simulation results applied to real images are displayed in (see Figs. 3, 4, and 5) below represent the detection of the points of interest by the different detectors (Harris, SURF, SIFT, and FAST) without modifying the threshold of uniqueness. Then we modify the threshold of uniqueness to find an optimal value at which the eight weak points can be detected with an average error equal to 1 pixel (Fig. 6). The different results are represented in the following figure:

After changing the detector uniqueness threshold (SURF) to obtain the eight points to calculate F, we then apply the one on the stereo images with different changes as follows: (1) lighting, (2) rotation, and (3) moving objects in the scene.

We apply our technique (Robust detectors (SIFT, ASIFT, SURF with RANSAC) to the facial image dataset (ORL) [36] and also obtain better results for the identification of the face by different changes, as shown in the figures below (Figs. 7, 8, and 9).

We can see from the table that the number of characteristic points and the number of inliers detected by Harris, FAST, and SURF are lower than with the SIFT method. The simulation results show that this method shows better results compared to the others in terms of projection errors with different variations of the scene (see Fig. 10).

Fig. 11 shows a comparison of our method and the other methods in terms of processing speed for estimating F.

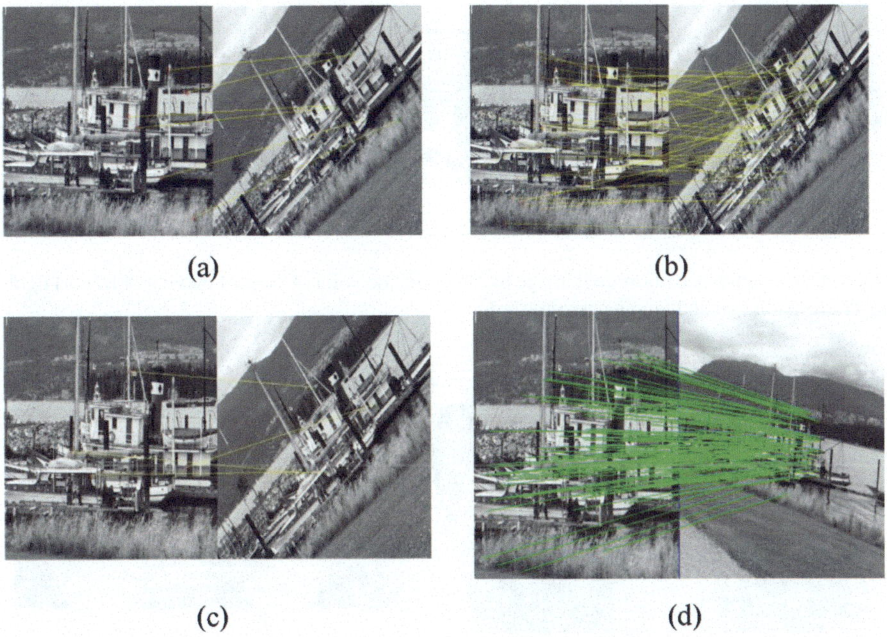

Fig. 4 Correspondence of original image in rotation: Harris (a), SURF (b), FAST (c), and SIFT (d)

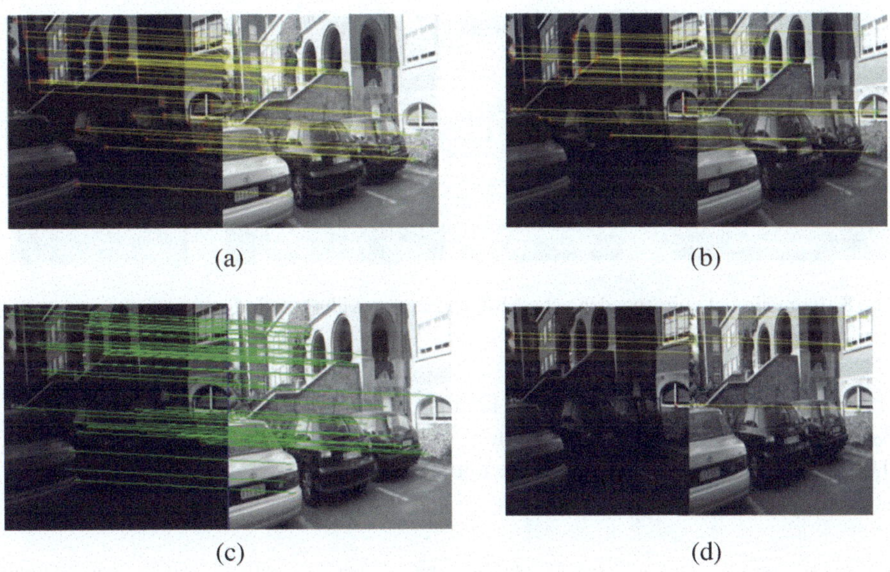

Fig. 5 Correspondence of original image by change in lighting: SURF (a), Harris (b), SIFT (c), and FAST (d)

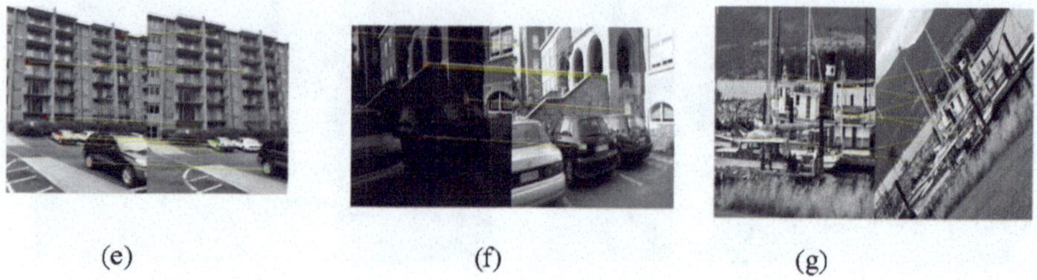


**Fig. 6** Correspondence of original image by changing the point of view of moving object (e) lighting (f), rotation (g), and uniqueness threshold

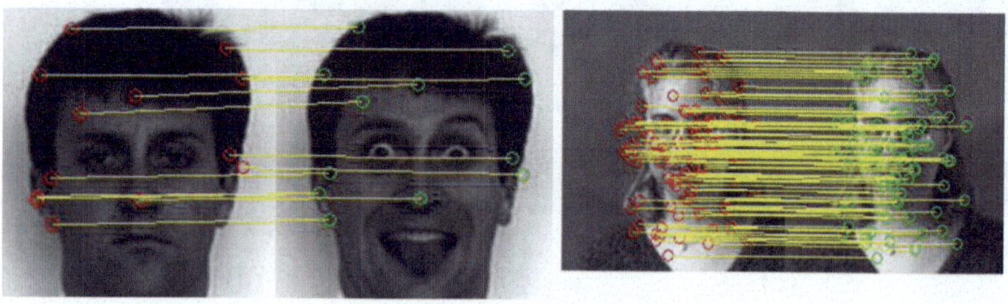

**Fig. 7** Extraction of correspondence by SIFT detectors with RANSAC

**Fig. 8** Extraction of correspondence by ASIFT detectors with RANSAC

The proposed method gives good results in terms of calculation time of the fundamental matrix with different variations of the scene that generally does not exceed 0.008 s; it can be applied to real-time applications.

## 6 Conclusions

This paper presents two methods based on a comparison of detected points (descriptor), the correct correspondence, the time to estimate the fundamental matrix F, the projection error, recognition rate, and authentication time. The classification of points is done

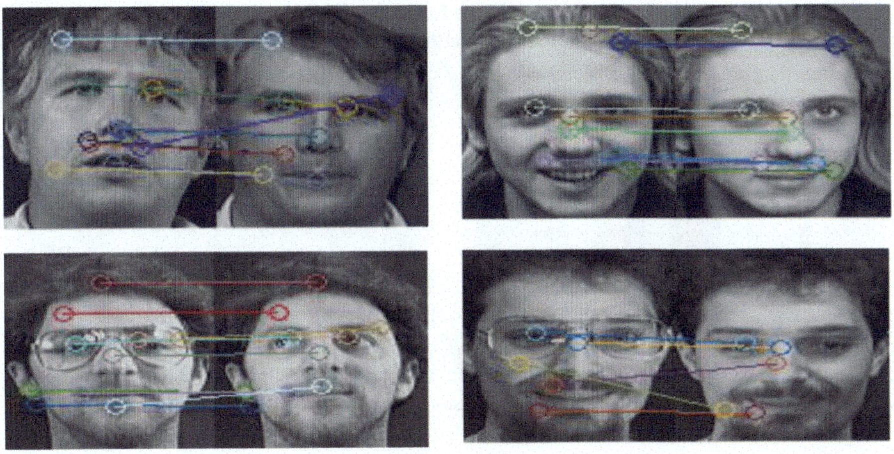

**Fig. 9** Extraction of correspondence by SURF detectors with RANSAC

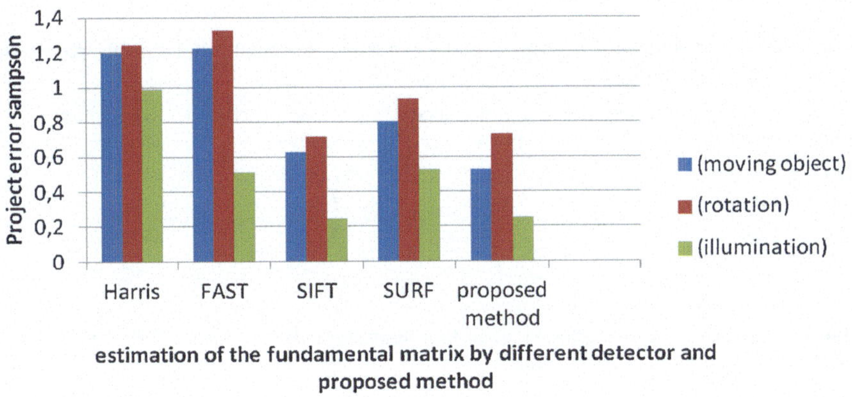


**Fig. 10** Error project Sampson

by changing the weighting function to "multi-level". The primary technique combined the different descriptors of the SIFT, SURF, FAST, and Harris methods with the RANSAC algorithm using eight standardized points, which makes it possible to obtain a very large number of descriptors and correspondence. On the other hand, this technique, when applied to a face database (ORL), gives good results in terms of recognition rate and authentication time. The critical disadvantage of the estimation of F is related to the time of calculation, which exceeds 1 s.

The second technique based on the variable uniqueness threshold SURF descriptor is associated with the eight-point standardized M-estimator algorithm to obtain F at different levels using a weighting function. The average error of the estimate of F is very important in comparison to the previous method. But in terms of the

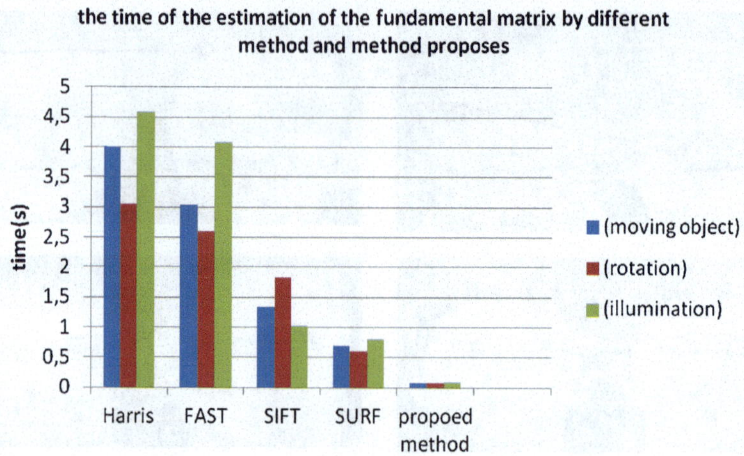


**Fig. 11** The different detectors combined with the RANSAC algorithm, the eight standardized points and the proposed method

**Table 1** Results of simulation in terms of entity extraction, key point, similar points, recognition rate, and average processing time

| Method | Average processing time (s) | Average of detected descriptors | Average of correct matches | Recognition accuracy (%) |
|---|---|---|---|---|
| SIFT | 0.78 | 40 | 37.6 | 94 |
| ASIFT | 1.78 | 60 | 57 | 95 |
| SURF | 0.69 | 20 | 19.12 | 95.6 |

**Table 2** Comparison of the performance of four algorithms (Harris, FAST, SIFT and SURF) in terms of key points and match selection

| | Without modification | | | | Wit threshold modification +SURF |
|---|---|---|---|---|---|
| the detectors | Harris (in motion) (rotation) (lighting) | FAST (in motion) (rotation) (lighting) | SIFT (in motion) (rotation) (lighting) | SURF (in motion) (rotation) (lighting) | (in motion) (lighting) (rotation) |
| Kypnt1 | 309 | 449 | 54 | 423 | 82 |
| | 273 | 112 | 24 | 152 | 18 |
| | 264 | 76 | 16 | 111 | 43 |
| Kypnt2 | 328 | 512 | 480 | 439 | 63 |
| | 271 | 110 | 142 | 126 | 21 |
| | 325 | 91 | 142 | 88 | 35 |
| Inliers | 40 | 54 | 153 | 150 | 8 |
| | 26 | 24 | 105 | 73 | 8 |
| | 32 | 16 | 86 | 46 | 8 |

estimation time of F, it gives good results from different points of view. In conclusion, this strategy can be used for real-time applications in, for example, moving object tracking and face authentication.

## References


1. Xu, G., & Zhang, Z. (2013). *Epipolar geometry in stereo, motion and object recognition: A unified approach* (Vol. 6). Berlin: Springer Science & Business Media.
2. Luong, Q. T., & Faugeras, O. (1997). Self-calibration of a moving camera from point correspondences and fundamental matries/. *International Journal Computer Vision, 22*(3), 261–289.
3. Golparvar-Fard, M., Peña-Mora, F., & Savarese, S. (2009). D4AR-A 4-dimensional augmented reality model for automating construction progress data collection, processing and communication. *Journal of Information Technology in Construction (ITcon), 14*, 129–153. Special issue next generation construction IT: technology foresight, future studies, road mapping, and scenario planning.
4. Golparvar-Fard, M., Pena-Mora, F., & Savarese, S. (2010). D4AR – 4 dimensional augmented reality – Tools for automated remote progress tracking and support of decision-enabling tasks in the AEC/FM industry. In *Int. Conf. on Innovations in AEC, Jun 2010.*
5. Teizer, J., & Vela, P. A. (2009). Personnel tracking on construction sites using video cameras. *Advanced Engineering Informatics, 23*(4), 452–462.
6. Wang, H., et al. (2009). Evaluation of local spatio-temporal features for action recognition. In *BMVC 2009-British machine vision conference*. Guildford: BMVA Press.
7. Tomasi, C., & Kanade, T. (1991). Detection and Tracking of Point Features Carnegie Mellon Univ. Tech. Rep., 91–132.
8. Lowe, D. G. (1999). Object recognition from local scale-invariant features. *Proceeding of International Conference on Computer Vision, 2*, 1150–1157.
9. Bay, H., Ess, A., Tuytelaars, T., & Van Gool, L. (2006). SURF: Speeded up robust features. *Computer Vision – ECCV, 2006*(3951), 404–417.
10. Kumar, P., Henikoff, S., & Pauline, C. N. (2009). Predicting the effects of coding non-synonymous variants on protein function using the SIFT algorithm. *Nature Protocols, 4*(7), 1073.
11. Snavely, N., Seitz, S. M., & Szeliski, R. (2008). Modeling the world from internet photo collections. *International Journal of Computer Vision, 80*(2), 189–210.
12. Huang, J., Lai, S., & Cheng, C. (2007). Robust fundamental matrix estimation with accurate outlier detection. *Journal of Information Science and Engineering, 23*, 1215–1225.
13. Cyganek, B., & Siebert, J. P. (2011). *An introduction to 3D computer vision techniques and algorithms*. Hoboken, NJ: Wiley.
14. Zhang, Z. (1998). Determining epipolar geometry and its uncertainty: A review. *International Journal of Computer Vision, 27*(2), 161–195.
15. Szeliski, R. (2010). *Computer vision: Algorithms and applications*. Berlin: Springer Science & Business Media.
16. Xiao, C.-B., Feng, D.-Z., & Yuan, M.-D. (2018). An efficient fundamental matrix estimation method for wide baseline images. *Pattern Analysis and Applications, 21*(1), 35–44.
17. Jog, G. M., Fathi, H., & Brilakis, I. (2011). Automated computation of the fundamental matrix for vision based construction site applications. *Advanced Engineering Informatics, 25*(4), 725–735.
18. Zhang, K., Li, X. Z., & Zhang, J. X. (2014). A robust point-matching algorithm for remote sensing image registration. *IEEE Geoscience and Remote Sensing Letters, 11*(2), 469–473.
19. Bay, H., et al. (2008). Speeded-up robust features (SURF). *Computer Vision and Image Understanding, 110*(3), 346–359.

20. Chater, A., & LAsfar, A. (2018). Reconnaissance d'expression faciale basée sur la fusion de PCA, LBP et LPQ avec SVM. In *Conférence Internationale CITISI'18*.
21. Rublee, E., et al. (2011). ORB: An efficient alternative to SIFT or SURF. In *Computer Vision (ICCV), 2011 IEEE international conference on*. Piscataway, NJ: IEEE.
22. Yao, G., et al. (2018). Robust Harris corner matching based on the quasi-homography transform and self-adaptive window for wide-baseline stereo images. *IEEE Transactions on Geoscience and Remote Sensing, 56*(1), 559–574.
23. Ballerini, M., et al. (2008). Interaction ruling animal collective behavior depends on topological rather than metric distance: Evidence from a field study. *Proceedings of the National Academy of Sciences, 105*(4), 1232–1237.
24. Zhang, Z. (2000). A flexible new technique for camera calibration. *IEEE Transactions on Pattern Analysis and Machine Intelligence, 22*(11), 1330–1334.
25. Chater, A., & Lasfar, A. (2018). Detection of image descriptors and modification of the weighting function for the estimation of the fundamental matrix using robust methods. *Journal of Engineering and Applied Sciences, 13*, 1835–1843.
26. Li, Y., Velipasalar, S., & Gursoy, M. C. (2013). An improved evolutionary algorithm for fundamental matrix estimation. In *Advanced video and signal based surveillance (AVSS), 2013 10th IEEE international conference on*. Piscataway, NJ: IEEE.
27. Hartley, R., & Zisserman, A. (2003). *Multiple view geometry in computer vision*. Cambridge university press.
28. Hartley, R. (1995). In defence of the eight point algorithm. In *Proc. 5th Int. Conf. Comput. Vision* (pp. 1064–1070). Boston, MA: IEEE Computer Science Press.
29. Chater, A., & Lasfar, A. (2019). Robust Harris detector corresponding and calculates the projection error using the modification of the weighting function. *International Journal of Machine Learning and Computing, 9*(1), 62–66.
30. Hartley, R., & Zisserman, A. (2000). *Multiple view geometry in computer vision.*, Second ed (p. 2003). Cambridge: Cambridge University Press.
31. Armangué, X., & Salvi, J. (2003). Overall view regarding fundamental matrix estimation. *Image and Vision Computing, 1*, 205–220.
32. Raguram, R., Frahm, J.-M., & Pollefeys, M. (2008). A comparative analysis of RANSAC techniques leading to adaptive real-time random sample consensus. In *European conference on computer vision*. Berlin: Springer.
33. Zhou, F., Zhong, C., & Zheng, Q. (2015). Method for fundamental matrix estimation combined with feature lines. *Neurocomputing, 160*, 300–307.
34. Wu, B., Zhang, Y., & Zhu, Q. (2011). A triangulation-based hierarchical image matching method for wide-baseline images. *Photogrammetric Engineering & Remote Sensing, 77*(7), 695–708.
35. Fathi, H., & Brilakis, I. (2011). Automated sparse 3D point cloud generation of infrastructure using its distinctive visual features. *Journal of Advanced Engineering Informatics, 25*(4), 760–770.
36. *AT&T Database of faces 'ORL face database'*. Cambridge: AT&T Laboratories. http://www.cl.cam.ac.uk/Research/DTG/attarchive/facedatabase.html.
37. Chater, A., & Lasfar, A. (2019). Comparison of robust methods for extracting descriptors and facial matching. In *2019 international conference on wireless technologies, embedded and intelligent systems (WITS)* (pp. 1–4). Piscataway, NJ: IEEE.

# Identification of Persons Using Stereo Image Pairs and Perturbation Functions

**Sergey Vyatkin, Alexander Romanyuk, Sergey Pavlov, and Oksana Romanyuk**

## 1 Introduction

For a considerable period, image analysis and computer graphics have evolved independently without finding common ground. This is not surprising: in fact, computer graphics and image analysis are two symmetrical methods of processing visual information. Computer graphics operates with formal descriptions of objects to create their visual images. On the other hand, image analysis systems work with images to produce formalized object models in one form or another. Recently, there has been a trend towards convergence and mutual integration of computer graphics and image analysis. This is primarily due to the development of virtual reality systems, such as computer games, various training systems, simulators for pilots, drivers and astronauts.

In theory, virtual reality is an absolute interface between a person and a computer, which uses all or almost all systems of human interaction with the outside world. In this sense, the user of the virtual reality system experiences the sensation of interaction with some model environment that has the features of the real world, without contact with it. The model environment and its perception are usually fully modeled on a computer. One of the most important problems in the development of a virtual reality system is the construction of a model environment. The world in which the interaction between the system and the user takes place. The modeling environment may consist of objects created entirely in the computer with the hands of the designer, and of the objects "digitized" with real. In other words, descriptions

S. Vyatkin (✉)
Institute of Automation and Electrometry of the Siberian Branch of the Russian Academy of Sciences, Novosibirsk, Russian Federation
e-mail: sivser@mail.ru

A. Romanyuk · S. Pavlov · O. Romanyuk
Vinnytsia National Technical University, Vinnytsia, Ukraine
e-mail: romanyuk@vntu.edu.ua; psv@vntu.vinnica.ua; oroman@vntu.edu.ua



S. Bilan et al. (eds.), *Biometric Identification Technologies Based on Modern Data Mining Methods*, https://doi.org/10.1007/978-3-030-48378-4_8

of objects in the form of their geometric models can be obtained by various methods of analysis of real objects. One of the most common methods of obtaining such a description is based on the analysis of image pairs.

The field of study of this work is a set of solutions to the problem of determining the spatial parameters of the object by the stereo pair of its images. Obtaining three-dimensional models of real objects by analyzing pairs of their images is necessary in order to add virtual objects to them for subsequent use in virtual reality systems.

The aim of the work is to develop a method and implement a program based on it to obtain a real-time three-dimensional model of the visible part of the surface of the scene observed by two video cameras.

## 2 Restoration of Three-Dimensional Coordinates of a Scene on Stereo Pairs of Images

Geometry reconstruction refers to restoring the elevation map of the entire scene. Height map—a two-dimensional array ($x$, $y$), in which each cell is the value of the coordinate Z. Hence, our goal is to get a map of the heights of the scene by binocular vision.

Consider a situation where two cameras at different points register the same scene. A pair of images obtained in this case is called a stereo pair. Let us first turn to the simplest case. Let the same cameras are arranged so that their optical axes are parallel, and the line passing through the optical centers is perpendicular to the optical axes (this line is called the baseline, and its segment enclosed between the optical centers—the base). Put the length of the base equal to $B$. Choose a global coordinate system whose origin $O$ is located on the baseline in the middle between the optical centers of the cameras, the OZ axis is parallel to the optical axes, and the OX axis is directed along the baseline.

Let the coordinates in the camera image planes coincide with the main points $U_0 = V_0 = 0$, and the coordinate units in the global system and in the camera image planes are the same ($w = h = 1$).

Select a point with global coordinates ($X, Y, Z$). The coordinates of its projection in the image plane of the first (left) camera are denoted by $(x', y')$, and in the image plane of the second (right) camera by $(x'', y'')$. Projections of the same point $M$in the planes of images of different cameras are called conjugate points. It is not difficult to verify that

$$x' = f\left(X + b/2\right)/Z,\ x'' = f\left(X - b/2\right)/Z,\ y' = y'' = fY/Z$$

Note that in the direction perpendicular to the direction of the baseline, the coordinates of the conjugate points ($y$-coordinates) are the same. This fact is of great importance in the automated search for conjugate points on the stereo pair, allowing to significantly reducing the size of the search area. From the first two relations it follows that

$$Z = fb / \left(x' - x''\right)$$

This means that, knowing the geometry of the shooting and measuring the coordinates of the projections of the same point in the planes of the camera image, you can calculate the depth (coordinate *Z*) of this point. Moreover, thus generated correlations enable calculating all the three-dimensional coordinates of the point:

$$X = b\frac{\left(x' + x''\right)}{2\left(x' - x''\right)}, \quad Y = b\frac{\left(y' + y''\right)}{2\left(x' - x''\right)},$$

The difference $d = x' - x''$ is called disparity. From the formulas, it becomes obvious that errors in the coordinates of the projections are more affected by low variance and, consequently, the distances to distant objects are measured less accurately than to close ones. On the other hand, at a fixed range, the variance is proportional to the size of the base; hence, the accuracy of measurements increases with increasing base. Further, we will see, however, that the increase in the base can lead to errors that are not compensated by the increase in the accuracy of measurements.

From all the above it follows that to estimate the three-dimensional coordinates of a point on the stereo pair:

1. it is necessary to know the internal parameters of the cameras (the task of their calibration),
2. it is necessary to know the parameters of the relative position of the cameras (the problem of mutual orientation),
3. it is necessary to find and determine the coordinates of the corresponding conjugate points on the images (the task of finding conjugate points).

## 3 Conditions and Assumptions

From the theory described above it is clear that the main difficulty of the development described in this chapter of the algorithm is the correct choice of the mechanism for finding the corresponding points. Therefore, additional conditions and assumptions are introduced that limit the search space and thus reduce the likelihood of false matches.

Epipolar restriction. As mentioned, the matching space is narrowed to one-dimensional (along the epipolar), which significantly increases the efficiency of matching and reduces the probability of error. In the case of a standard binocular configuration epipoles, correspond to the scan lines of the image, which further facilitates the work of search algorithms.

The smoothness condition states that scenes can only contain smooth and opaque objects. Thus, for the depth (and, accordingly, the disparity) along the surfaces of objects, the condition of continuity is fulfilled. Namely, if the features of p1 and ql on the target and offset image are corresponding, then any feature of p2 from the neighborhood of p1 on the target image can correspond only to such a feature of q2 on the offset image that the difference between the norms of the disparity pairs (p1, ql) and (p2, q2) does not exceed the modulo in advance of the given small value. It is obvious that the smoothness condition is not satisfied near the object boundaries. Therefore, many algorithms determine the boundaries of objects (and, accordingly, the area of the barrier) precisely by a sharp change in the disparity.

The condition of limitation on the value of the disparity states that for any pair of corresponding features on the target and offset image, the disparity does not exceed the norm of the specified value. Thus, only objects that lie at a certain distance from the camera are considered.

Condition of orderliness states that the order of the corresponding features on epipolar does not change. Although this restriction does not apply to narrow objects that are close to the camera, its awareness has allowed creating a number of effective algorithms.

The uniqueness condition states that any feature in the key image can only match one feature in the offset image. Thus, transparent objects and areas lying along the direction of view of the key camera are not considered.

The photometric compatibility condition states that for any pair of corresponding singularities (pl,ql) and (p2,q2) on the target and offset image of the intensity difference, the absolute intensity difference does not exceed the modulo pre-set value. Thus, it is assumed that the corresponding points on the target and offset image have approximately the same intensity.

## 4 System Calibration

The task of calibration of stereo cameras includes the calculation of parameters such as focal lengths, the relative position of the cameras in space, the parameters of optical curvature, which have all cameras, the parameters of brightness distortion, etc. Calibration methods are divided into photogrammetric [1–6] and self-calibration [7–11]:

Photogrammetric calibration. It is based on the observation of an object whose geometry is known with very good accuracy. A calibration object usually consists of two or three planes orthogonal to each other. Information about its geometry and the elevation maps taken by the cameras are used to determine the system parameters.

Self-calibration. This technology does not use any calibration objects. Camera movement is used in a rigid static scene, which makes it possible to determine the internal parameters of the cameras. If the internal parameters of the cameras are fixed, it is possible to determine all the parameters of the scene when moving. This approach is very promising, but it does not always give the right results, because a very large set of parameters needs to be defined.

To solve the problem calibration can be carried out by any photogrammetric method. The next chapter will offer one of the easiest to apply to the task.

## 5 Match Search Task

Studies of the mechanisms of stereopsis and binocular optical systems have been carried out for many years, and different authors have developed a significant number of algorithms that allow solving the problem of finding corresponding features with one or another efficiency. In general, according to the type of processed features, all algorithms for constructing depth maps can be divided into two large groups:

The algorithms for geometric features operate on the geometric features, such as angles, straight lines, etc., highlighted in the image with the help of filters. The geometrical features with the most similar parameters are declared offset. The advantages of the algorithms for geometric features are low probability of false matches and high accuracy of finding the disparity (modern filters easily find geometric features at the subpixel level). The main drawback of this kind of algorithms can be considered the relative sparsity of the resulting depth maps.

The algorithms for regions operate on pixels and their intensities. A feature in the general sense is considered a pixel, the parameter features-intensity. Such algorithms were first developed for processing aerial photography data of undeveloped areas. Such data are characterized primarily by the smoothness of the surface to be restored and the lack of a sufficient number of geometric features. In addition, the condition of the "plane of the world" is fulfilled—in local areas the target surface can be considered flat. The algorithms for the regions allow us to obtain sufficiently dense depth maps (ideally, the matching problem is solved for each visible pixel on the target and offset image). The disadvantages of such algorithms arise from the orientation to the "plane of the world" condition (as a consequence, unstable work in the areas of depth gap) and the use of only pixel intensity as parameters (as a consequence, unstable work with different camera parameters and in cases of lighting changes).

It is clear that to solve the problem of finding a match for areas, you need to set the criterion of "similarity" of pixels, which allows you to select the only one corresponding to the fixed pixel on the target image among the set of candidate pixels on the offset image. It is obvious that simple criteria like the intensity difference module are not suitable. Therefore, it is common to do the following: the pixels to be tested on the target and offset images are surrounded by small windows of a fixed size and determine the "similarity" of the windows based on the calculation of the correlation coefficient. Checking windows instead of directly checking the intensity allows you to avoid errors related to noise and local changes in illumination. On the other hand, in the case of choosing a "bad" correlation criterion, the use of windows leads to unstable operation of the algorithm in the areas of sharp changes in depth (they do not meet the condition of the "plane of the world").

The most well-known algorithms for regions can be divided into three classes:

Local algorithms. In this case, the matching problem is solved independently for individual pairs of points (pixels) on the target and offset image. When solving the matching problem for each fixed pair of points, the results of the solutions for the other pairs are not taken into account. Each pair of corresponding points is searched independently and in parallel with other pairs.

Search algorithms in the epipolar space. In this case, the matching problem is solved for epipolar pairs on the key and offset image. Each pair of epipolar forms an epipolar space, and although the points within the space are compared independently based on the correlation criterion, the result of solving the matching problem for a given pair of points will be used to solve the problem for the next pairs of points.

The global algorithms compare the images. In this case, the target image and the offset image are compared as a whole, and the result of solving the matching problem is determined simultaneously for all points of the target image. It should be noted that although binocular visual systems predominate in nature, developers of algorithms for calculating depth maps usually use a larger number of cameras. Redundancy of information (three images instead of two), appearing in this case, allows to cope more effectively with the ambiguities characteristic of the stereopsis algorithms.

Criteria for the pixels:

Among the criteria for comparing pixels, two large groups can be distinguished. These are parametric and nonparametric criteria. Parametric tests explicitly use the values of the intensities of the pixels in the window.

## 6 Parametric Tests

Among the parametric criteria, the most widely used is the sum of squares of SSD intensity differences. For the window of radius R, the sum of the squares of the intensity differences is defined as

$$SSD(x_0,y_0,d)=\sum_{x=-R}^{R}\sum_{y=-R}^{R}\left(I_l(x+x_0,y+y_0)-I_r(x+x_0,y+y_0)\right)^2$$

The SSD criterion uses the order of pixels in the window (which allows you to successfully discard symmetric areas) and integer calculations. It can be successfully implemented in linear form, so its complexity will be O(width*height*(disparity_map-disparity_map)). You can also use its modification, which is insured against brightness errors of video cameras, when the average brightness may differ from the cameras on constant or constant times:

$$SSD(x_0,y_0,d)=\frac{\sum_{x=-R}^{R}\sum_{y=-R}^{R}\left(\left(I_l(x+x_0,y+y_0)-\overline{I}_l\right)-\left(I_r(x+x_0,y+y_0)-\overline{I}_r\right)\right)^2}{\sqrt{\sum_{x=-R}^{R}\sum_{y=-R}^{R}\left(\left(I_l(x+x_0,y+y_0)-\overline{I}_l\right)*\sum_{x=-R}^{R}\sum_{y=-R}^{R}\left(I_r(x+x_0,y+y_0)-\overline{I}_r\right)\right)}}$$

## 7 Nonparametric Criteria

Nonparametric criteria do not directly use intensity values, but try to somehow track the topology of the space that falls into the window. Among nonparametric criteria, the census criterion is the most common [12, 13]:

$$SSD(x_0,y_0,d)=\sum_{x=-R}^{R}\sum_{y=-R}^{R}\left(I_l\left(x+x_0,y+y_0\right)>I_l\left(x_0,y_0\right)\right)\wedge$$
$$\left(I_r\left(x+x_0,y+y_0\right)>I_r\left(x_0,y_0\right)\right)$$

Here operation ">" takes the values 1 if the left operand is greater than the right one and 0 otherwise. The "^" operation takes the values 1 if the left operand is not right and 0 otherwise.

This criterion carries protection against glare and brightness errors of video cameras, when the average brightness may differ from the cameras on constant or constant times. However, it has big disadvantage — complexity O(R*R*width*height*(disparity_map-disparity_map)).

In addition, the authors [13] proposed an interesting criterion that can be used when the preliminary segmentation of images on the area is carried out. Each area has its own set of parameters that define its distortion. These parameters are found using the correlation coefficient. Two methods of cross-correlation and morphological correlation are considered. In both cases, the problem of finding parameter corrections is reduced to the problem of generalized eigenvalues. Unfortunately, this criterion is too time-consuming to use in solving the problem.

For calibration, it was decided to use the photogrammetric method, since the parameters of the cameras are assumed unchanged and it is enough to determine them once. To find matches, it was decided to use the search algorithm in the epistolary space as fast and efficient enough. During the comparative test of SSD and census criteria, SSD was selected.

The simplest binocular system is when the optical axes of the cameras are assumed parallel, the cameras are fixed, the observation planes coincide, and the internal parameters of the cameras coincide. Due to the simplicity of the calibration implementation, an approach similar to that described in [1] was chosen. Several shots of a flat object with points on the surface are taken. Points are plotted in a certain order and knowledge of this order is used to find matches on stereo images. After finding the matches, the knowledge of the distance between the points is used, a system of equations for the internal and external parameters of the scene is compiled and the least squares method is solved, minimizing errors in the measurement of these parameters.

It is also possible to simplify the method—the definition of not all parameters, but only the focal length and the size of the base. Alternatively, more precisely, you need to find the product of these two quantities to use them to obtain Z:

$$Z=fb/\left(x'-x''\right)$$

The authors [14] conducted extensive testing of matching algorithms on pairs of images and found that the fastest is the sum-of-squared-differences algorithm—the difference of the sum of squares. In addition, indeed this is the most simple. This algorithm can be briefly described as the follows:

1. Calculate the price of matching a pair of pixels as the square of the pixel intensity difference with the given disparity.
2. The summation of the prices on the window with a specified constant size.
3. The disparity is selected from at least all values for each pixel.

Unfortunately, this algorithm shows too little stability because many conditions are not checked in it. Therefore, the following algorithm with the use of dynamic programming was chosen.

Consider a standard binocular system and an epipolar plane uniquely defined by the optical centers of the cameras and an arbitrary point in space. A couple of epipolar formed by the intersection of the epipolar plane and the picture planes of the target and offset of the image determines a discrete space, called an epipolar search space. Consider a couple of these epipolar (el, er) as the coordinate axes. The elements of the discrete epipolar search space B(i,j) are the values of the correspondence criterion for the I-th point el(i) epipolar on the target image and the j-th point er(j) epipolar on the offset image. The smaller the value of the criterion, the more likely it is that the points el(i) and er(j) form a corresponding pair. To solve the matching problem, let us move to a more convenient form of representation of the search space. For a standard binocular system, the pairs of epipolar will correspond to the pairs of scan lines of the target and offset images. Consider a couple of these epipolar (el(y) er(y)). We introduce a discrete search space as follows: on one axis, we lay the values of the coordinate x along the epipolar el(y) on the target image, and on the other—the values of the disparity. The elements of the space ESSP (x,d,y) will be the match criterion values for the point(x) of the El (y) epipolar on the target image and the point (x-d) of the epipolar on the offset image. The smaller the value of the criterion, the more likely it is that the points el(x, y) and er(x-d,y) form a corresponding pair.

Consider all possible paths from the lower left corner to the right edge of the search space ESSP(x,d). Obviously, each of these routes is a solution to the matching problem for all epipolar points on the target image. If there are restrictions on the "form" of the path, using the method of dynamic programming, we can quickly choose from a variety of routes optimal in terms of cost. Thus, by correctly setting the cost and constraint function, we can effectively solve the matching problem for all points of the epipolar pair at once. Obviously, the desired optimal route in the search space cannot be of arbitrary shape. Make important assumptions:

(a) A scene consists of objects that may exist in physical reality. Therefore, we consider only one-connected routes.

(b) All objects present in the scene are opaque. Thus, the visible point can correspond to only one point. If it corresponds to more than one point, it is considered to be blocked.
(c) The condition of order is fulfilled. From these assumptions, it immediately follows that the desired path cannot change direction, i.e., for the route point $(i,j)$ only the point $(i + l, j)$, $(i, j + 1)$, or $(i + 1, j + 1)$ can be next. This fact allows you to successfully apply the method of dynamic programming.

We introduce the cost function as the arithmetic mean of the cost of the pixels for which we have already built a route. Moreover, we will look for a route that minimizes this function. The transformation is specified by the rule:

$$m = i - j, n = j$$

In such a space, rule (C) will be reformulated: for a route point (m,n), only the point (m+1, n), (m, n+1), or (m−1, n+1) can be next. To build the optimal route we will use a modified Dijkstra algorithm. Below is its description:

Task. It is known that all prices are non-negative. Find the minimum cost function path from the lower boundary of the space to the top.

We put the entire lower bound of the space in the list of points sorted by the cost function, for which the optimal path has already been found, and start the cycle.

1. We get from the list the first point. Remove it from the list.
2. Check if it lies on the upper border of the exit cycle to point 4.
3. For each of the three nearest points, we determine whether they were selected from the list earlier. If it happened, go to the beginning of the cycle in point 2. Otherwise, we calculate a new cost function for it, assign the current point as the "parent" for this neighboring point, and add the neighboring point to the list.
4. Restore the prescribed "parents" the desired path.

The hardest part of this algorithm is the constant sorting of the list. When you set it as a tree, it occurs in nLog (n) steps. The testing showed that of the point lying closer to the end of the list almost never contribute to the construction of the route. The introduction of an additional point to the algorithm is obvious: 2 (a) if the length of the list reaches the constant C2, and then delete all its elements following the number C1. (C1 < C2).

Constant value C1 was chosen to be equal to the width of the search space, i.e., (disparity_max-deparity_min). This decision is based on the peculiarities of filling the search space—if a poorly textured surface occurs in the scene; the space is filled with small values close to each other. It is from such areas that the most points fall into the list. Obviously, such areas are bounded by the width D of the search space. The value of C2 was chosen as 3C1. Testing has shown that adding this small check reduces the length of the sorted list by several times without making noticeable distortions.

## 8 The Method of Face Recognition

The method of face recognition with the use of perturbation functions and the set-theoretic operation of subtraction was presented in [15, 16].

A calibrated stereo pair (Fig. 1) is used for calculating 3D points on the face. Let us assume that we have two projective matrices $M_i$

$$\begin{pmatrix} u_i s_i \\ v_i s_i \\ s_i \end{pmatrix} = M_i \begin{pmatrix} X \\ Y \\ Z \\ 1 \end{pmatrix}$$

where *x*, *y*, *z* are three-dimensional coordinates of the point, ui and vi are their projections in the image i, and si is the scale factor. The stereo pair is characterized by the following parameters: the points of the image planes $E1 = (u1, v1)$ and $E2 = (u2, v2)$, and the point of the world coordinate system $P = (x, y, z)$.

Using the calibrated stereo pair for the face, we calculate the depth map by the correlation algorithm. In this work, we use an area-based algorithm with correlation of image intensity levels [17].

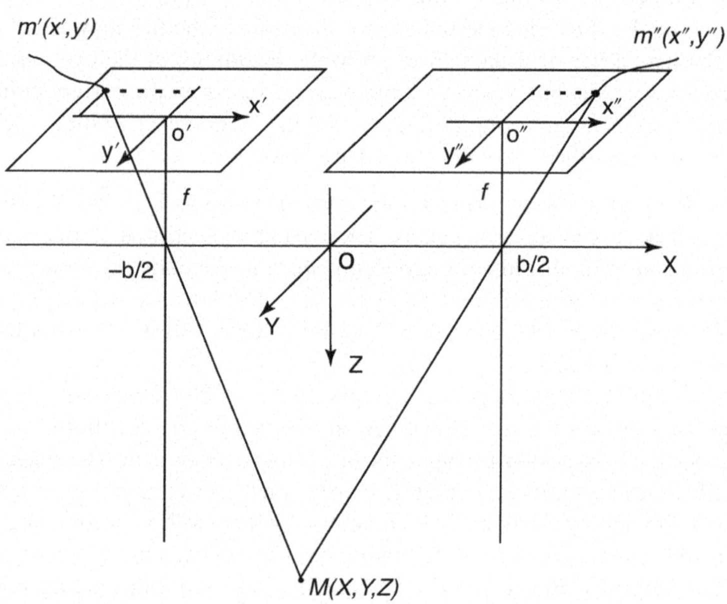


**Fig. 1** Stereo-pair

$$S = \frac{\sum_{i,j}\left(\left(I1(x+i,y+j)-\vec{I}1\right)-(I2(x+dx+i,y+dy+j)-\vec{I}2)\right)^2}{\sqrt{\left(\sum_{i,j}\left(I1(x+i,y+j)-\vec{I}1\right)^2\right)\left(\sum_{i,j}(I2(x+dx+i,y+dy+j)-\vec{I}2)^2\right)}}$$

Here $I1$ and $I2$ are the intensities of the left and right images, $\vec{I1}$ and $\vec{I2}$ are their mean values, dx and dy are the displacements along the epipolar line, and $s = \max(0.1 - c)$ is the correlation estimate.

There are two images of the stereo pair; scanning of these images provides information about the depth buffer (depth map) (Fig. 2).

In finding the perturbation peak, we calculate the characteristic size of the projection of the current interval, which is used as a basis for determining the detail level. For a larger interval, a rough approximation of the original function is taken. If a more detailed presentation is needed, then bilinear or bicubic interpolation of heights at the last detail level is performed. As a result, we obtain a 3D mask of the face (Fig. 3). Using three anthropomorphic masks, we construct a coordinate system that ensures a possibility of superposition of the tested masks (Fig. 4); finally, certain parts are cut off by a clipping plane for equalization of the volumes.

Applying the set-theoretic operation of subtraction

$$f_3 = \phi_t(f_1(x,y,z), f_2(x,y,z))$$
$$F_1 : f_1(x,y,z) \geq 0$$
$$F_2 : f_2(x,y,z) \geq 0$$

We determine the set of 3D points (voxels) belonging to the object $f3 = i(f1(x, y, z), f2(x, y, z))$, $\boldsymbol{F}1$: $f1(x, y, z)0$, $\boldsymbol{F}2$: $f2(x, y, z)0$. To find 3D points, voxelization of the remaining part of the volume after the subtraction is needed.

The smaller the number of voxels left, the greater the similarity of the tested objects.

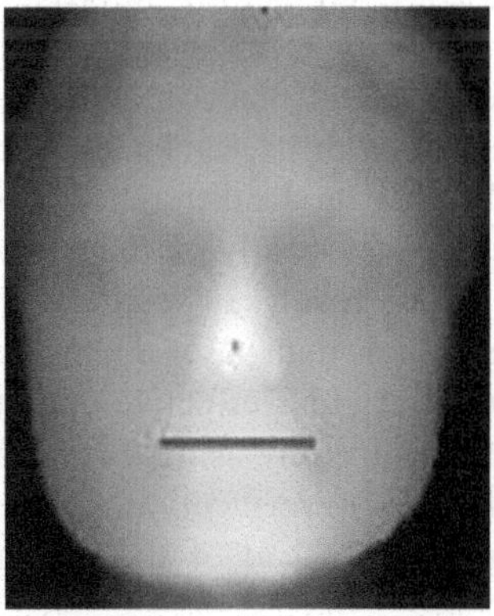

**Fig. 2** Depth map

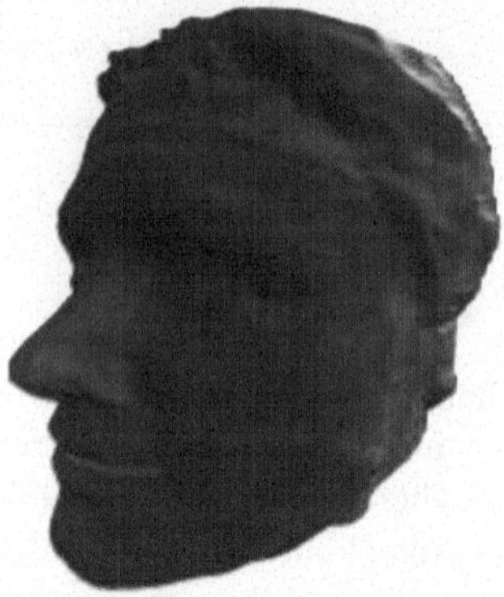

**Fig. 3** Reconstructed object

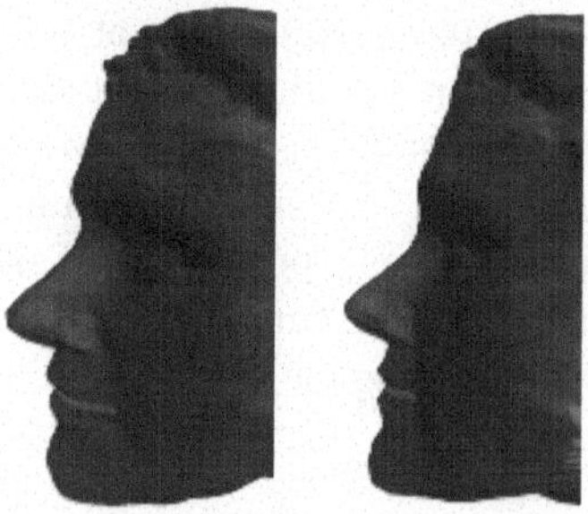

**Fig. 4** The test sample and the object from the database

## 9 Conclusions

During the experiments it was found that the effective size of the comparison window in the construction of the search space is determined by the resolution and size of the objects depicted on it.

Quality testing of several criteria was carried out by the sum of modules of differences, sum of squared differences, and the criterion of the census.

The first was discarded immediately, because its quality was almost the same as the second, but it is impossible to calculate a linear algorithm.

It was tested with a different set of permissions, using the SSD criterion. It is established that: on average, it takes approximately equal time to fill the search space with its values and search for the optimal path, the operating time depends linearly on the difference between the maximum and minimum disparities, the time depends linearly on the image area while maintaining the width-to-height ratio.

The results of testing the method are encouraging. Both virtual objects from available data bases and real persons were used. The 3D technology of face recognition provides effective operation; more than 98% of test objects were successfully recognized by using this method. Nevertheless, there are some factors that result in failure of verification. These factors can be classified into two groups: incorrect position ahead of the camera and interferences in data readout. The first class includes situations where only some part of the face is visible for the camera: the face is not directed toward the camera, the head is turned downward or to the left or right from the camera, the person is located too close to the camera, or the person goes away from the camera too fast after the beginning of verification (less than 1 s). The method operates successfully if the recognized object moves uniformly, but the camera fails to capture the observed object exactly in the case of its sudden acceleration.

It should be noted that observation of only some part of the face in the camera is not completely unacceptable because fragments can be successfully verified by using the geometric operation of intersection. The proposed method allows selective testing with the use of the geometric operation of intersection of a transparent cylinder or any other geometric shape with the surface.

Interferences of data readout occur if the facial expression is not neutral as required or if the headwear, mirror shades, or hair cover a major part of the face.

Advanced methods are capable of recognition based on different facial expressions.

Three-dimensional morphing is used for recognition in the proposed method.

If we compare 2D systems and the proposed 3D method of recognition, we can see that the false response probability in the first case is 0.12% and the false rejection probability is 9.8% for the recognition threshold being set at 70%. In the second case, the recognition threshold was set at 90%, and the method provided the false response probability of 0.004% and the false rejection probability of 0.1%.

In all tests performed simultaneously for both technologies with the use of the same images, the 3D technology of face recognition turned out to be more efficient than the 2D technology.

An example of 3D recognition methods is the well-known method of fitting for reconstructing the shape and parameters of the texture. This method is based on a system of linear equations. Recognition is performed on the basis of comparisons of the reconstructed shapes and texture of the image.

However, manual initialization is needed in the Face Identification by fitting a 3D Morphable Models method. The recognition time (approximately 1 min on the Pentium III processor with a frequency of 800 MHz) does not satisfy the requirements of most real systems.

As compared to previously available methods, the proposed method offers the following advantages: 3D morphing allows recognition of faces with different facial expressions; face identification on the basis of some part of the image is possible; texturing of the face surface is not needed; the method is completely automatic and fast (about 200 ms for one face image with a resolution of 640 × 480 pixels with the use of the Intel Core i7-2700K processor (8 MB cache memory, 3.90 GHz)), which is faster than the fitting method approximately by two orders of magnitude. The measurement error is no more than 0.8 mm (for each point of the 3D surface).

For real-time visualization, a binary method of searching for image elements with the use of graphics processing units adapted for calculating perturbation functions can be used. Therefore, a method of face recognition based on perturbation functions and the set-theoretic operation of subtraction is proposed.

Three-dimensional masks were used for face recognition. This method differs from available 3D methods by the fact that it involves not only all points of the surface in the recognition procedure, but also the volume of the tested mask. The method offers the following advantages: manual initialization of the process is not needed; three-dimensional morphing solves the problem of face recognition on the basis of different facial expressions; face recognition on the basis of only some part of the image is possible; face reconstruction is completely automated. The computation time is approximately 200 ms with a resolution of 640 × 480 pixels.

The method can be used in various situations where intellectual video monitoring of specially protected objects is needed: defense complex enterprises, heavily crowded areas, etc.

Further research areas:

1. Development of filters superimposed on the resulting elevation map.
2. Development of filters superimposed on the resulting three-dimensional search space ($x$, $d$, $y$) surface, after finding the surface heights.
3. Development of filters superimposed on the two-dimensional search space ($x$, $d$) before constructing the optimal route.
4. Development of criteria for comparison of pixels.
5. The development of methods of finding surface correspondences using a separate Y for each cost function, and the total for the entire scene.

The most promising works, in our opinion, are items 2 and 5. Since after constructing the search surface, we also have a lot of similar in price function surfaces, we can choose another, based on additional filters—for example, the median. Moreover, the development of a global cost function would help to reduce the number of errors in the construction of the search surface.

## References

1. Zhang, Z. (2000). A flexible new technique for camera calibration. *IEEE Transactions on Pattern Analysis and Machine Intelligence, 22*(11), 1330–1334. https://doi.org/10.1109/34.888718.
2. Faugeras, O., & Toscani, G. (1986). The calibration problem for stereo. In *Proceedings of the IEEE Conference on Computer Vision and Pattern Recognition* (pp. 15–20). Miami Beach, FL: IEEE.
3. Sturm, P. F. (2002). Critical motion sequences for the self-calibration of cameras and stereo systems with variable focal length. *Image and Vision Computing, 20*(5-6), 415–426. https://hal.inria.fr/inria-00525676/document.
4. Richard, I. (1997). Hartley. Self-calibration of stationary cameras. *International Journal of Computer Vision, 22*(1), 5–23.
5. Richard, I. (1992). Hartley estimation of relative camera positions for uncalibrated cameras. In *Proceeding ECCV '92 Proceedings of the Second European Conference on Computer Vision* (pp. 579–587). Berlin: Springer-Verlag.
6. Mendonça, P. R., & Cipolla, R. (1999, June). A simple technique for self-calibration. In *Proceedings. 1999 IEEE Computer Society Conference on Computer Vision and Pattern Recognition (Cat. No PR00149)* (Vol. 1, pp. 500-505). IEEE. https://doi.org/10.1109/CVPR.1999.786984

7. Kolesnik, M. *Direct recovery of the camera internal parameters using known angles.* https://www.researchgate.net/publication/228895886_Using_Angles_for_Internal_Camera_Calibration_and_Calibration_Update
8. Van Den Hengel, A., Brooks, M. J., Chojnacki, W., & Gawley, D. (2002, September). A New Constrained Parameter Estimator: Experiments in Fundamental Matrix Computation. In *BMVC* (pp. 1-9). https://www.researchgate.net/publication/221259813_A_New_Constrained_Parameter_Estimator_Experiments_in_Fundamental_Matrix_Computation
9. Maybank, S. J., & Faugeras, O. D. (1992). A theory of self-calibration of a moving camera. *The International Journal of Computer Vision, 8*(2), 123–152.
10. Karsten Mühlmann, Dennis Maier, Jürgen Hesser, Reinhard Männer. Calculating dense disparity maps from color stereo images, an efficient implementationJournal International Journal of Computer Vision archive 47 1–3, 2002, 79–88. Kluwer Academic Publishers Norwell, MA.
11. Zabih, R., & Woodfill, J. Non-parametric local transforms for computing visual correspondence. In *Proc. ECCV94. European conference on computer vision.* Berlin: Springer. http://www.cs.cornell.edu/~rdz/Papers/ZW-ECCV94.pdf.
12. Zabih, R., & Woodfill, J. (1996). A non-parametric approach to visual correspondence. In *Submitted to IEEE PAMI.* http://citeseerx.ist.psu.edu/viewdoc/download;jsessionid=B9CCD0F771FC8E08BBADAF88ED8AD147?doi=10.1.1.17.143&rep=rep1&type=pdf
13. Zheltov, S. Y., & Sibiryakov, A. V. (1998). Using An image preliminary segmentation for adaptive subpixel correlation. In *The 8-th International Conference on Computer Graphics and Visualizationics. Conference Proceedings. Moscow* (p. 7). https://www.academia.edu/2190839/Using_An_Image_Preliminary_Segmentation_For_Adaptive_Subpixel_Correlation.
14. Scharstein, D., & Szeliski, R. (2002). A taxonomy and evaluation of dense two-frame stereo correspondence algorithms. *International Journal of Computer Vision, 47*(1-3), 7–42. November 2001 Technical Report MSR-TR-2001-81.
15. Vyatkin, S. I. (2016). Method of face recognition with the use of scalar perturbation functions and set-theoretic operation of subtraction. *Optoelectronics, Instrumentation and Data Processing, 52*(1), 1–7. Allerton Press, Inc., 2016. ISSN 8756-6990.
16. Vyatkin, S. I., Romanyuk, S. A., Pavlov, S. V., & Necheporyk, M. L. (2018). Face identification algorithms and its using. *Modern Engineering and Innovative Technologies, 5*(Part 1), 111–115. https://doi.org/10.30890/2567-5273.2018-05-01-078.
17. Fua, P. (1993). A parallel stereo algorithm that produces dense depth maps and preserves image features. *Machine Vision Applications, 6*(1), 35–49.
18. Boykov, Y., Veksler, O., & Zabih, R. (1999). Fast approximate energy minimization via graph cuts. In *Seventh International Conference on Computer Vision (ICCV'99), Kerkyra, Greece* (pp. 377–384). Los Alamitos, CA: IEEE Computer Society.

# Identification of a Person by Palm Based on an Analysis of the Areas of the Inner Surface

**Mykola Bilan, Andrii Bilan, and Stepan Bilan**

## 1 Introduction

According to the analysis of the world market, biometric technologies bring great profit for different regions of the globe. By 2024, biometric technologies are expected to generate revenue of up to tens of billions of dollars per year. Almost every new smartphone has biometric technologies. Various characteristics are used as biometric parameters, which are divided into static (fingerprint, face shape, palm shape, vein pattern, ear shape, etc.) and dynamic (voice, gait, handwriting dynamics, etc.). Mainly geometric, brightness, and color forms represent static characteristics, and dynamic characteristics describe various quantities that change over time. Each of the biometric characteristics has its advantages and disadvantages. Moreover, for accurate and reliable biometric identification, several biometric parameters are simultaneously used. The best option is the simultaneous use of static and dynamic biometric characteristics.

Currently, experts are defining new biometric characteristics. Especially with the advent of new devices used by a large population. Now, for various smartphones, methods for biometric identification by the auricle, by the dynamics of working with the keyboard, identification of facial images and other characteristics have already been developed.

One of the simple methods of biometric identification is identification by the geometric shape of the hand. The parameters of the hand of each individual person are unique—the shape of the hand, the width, the length of the palm, individual

---

M. Bilan · A. Bilan
The Municipal Educational Institution Mayakskaya Secondary School, Mayak, Moldova

S. Bilan (✉)
State University of Infrastructure and Technology, Kyiv, Ukraine
e-mail: bilan_sm@gsuite.duit.edu.ua



S. Bilan et al. (eds.), *Biometric Identification Technologies Based on Modern Data Mining Methods*, https://doi.org/10.1007/978-3-030-48378-4_9

fingers, the folds of the phalanges on the back side, the pattern of blood vessels. In total, there are about 90 distinctive features.

A large number of palm biometric identification methods use such features as the width of the palm, the radius of the circle inscribed in the center of the palm, as well as the length of the fingers and the height of the hand [1–9]. The disadvantage of this approach is a the ability to easily create unauthorized copies of palm images.

There are many methods that use the geometry of the phalanx folds and the location of blood vessels [2, 7, 9]. Such methods use special devices to scan the palm and its fragments. The disadvantage is the possible damage to the shape of the palm.

To scan the palm of a person, many companies produce various devices that extract the totality of the necessary identification parameters. These devices can be easily acquired. Most of these devices have a design that requires the location of the hand with a given location. The location of the fingers is also taken into account.

The biometric identification of the geometric shape of the palm is comparable to the biometric identification of the fingerprint and is used in many access systems.

In this work, the task is to increase identification accuracy by analyzing the shape of the palm of the hand, which has an arbitrary location and is presented on a modified scale.

To solve this problem, the work proposes a system of relations of biometric parameters that the authors use to accurately describe the shape and area of the palm.

## 2 Description and Software Implementation of the Method

The geometric shape of the palm can be analyzed from the inside and outside. Biometric characteristics of both sides of the palm have significant differences. In this case, the area of the geometric shape of the palm and fingers are most often used. However, the area may be the same for different geometric shapes. Therefore, the analysis of the area of the palm should be carried out simultaneously with the analysis of the geometric shape.

The geometric shape of the palm can be described through a polygon approximating the outline of the palm. In this case, new additional quantitative characteristics appear, such as the length of the contour, the length of the segments between the vertices of the polygon, and the angles at the vertices.

Also, the geometric shape can be described by the areas of individual regions inside the image contour. To do this, the geometric shape of the inscribed circle or rectangle is selected, the area of which is simply calculated, and the remaining regions that do not belong to the inscribed figure have an area value that allows you to describe the geometric shape in the selected edge of the palm.

Since the methods of biometric identification by the geometric shape of the palm of the hand depend on devices, the images from the scanning device are formed clearly with well-defined borders and color scheme (Fig. 1) [2].

In most methods, an image with spread fingers is a prerequisite. Therefore, scanners contain the necessary devices for spreading fingers. However, there are methods that are enough to spread two fingers, as well as methods that do not take into

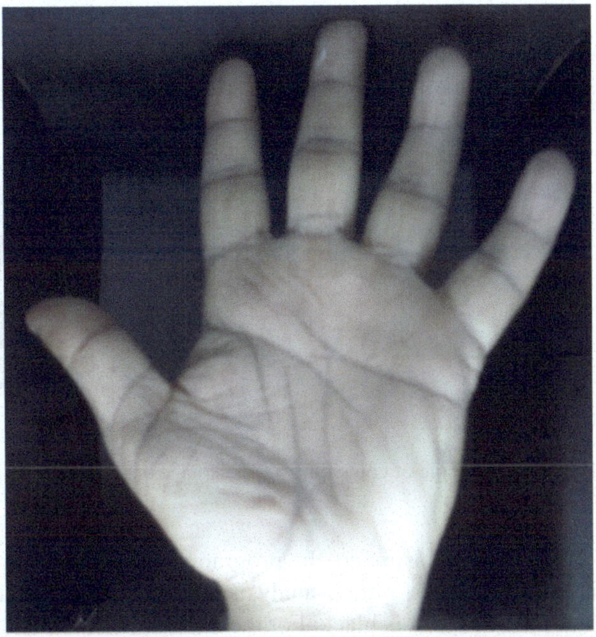

**Fig. 1** Image of the inner side of the palm obtained from the scanning device

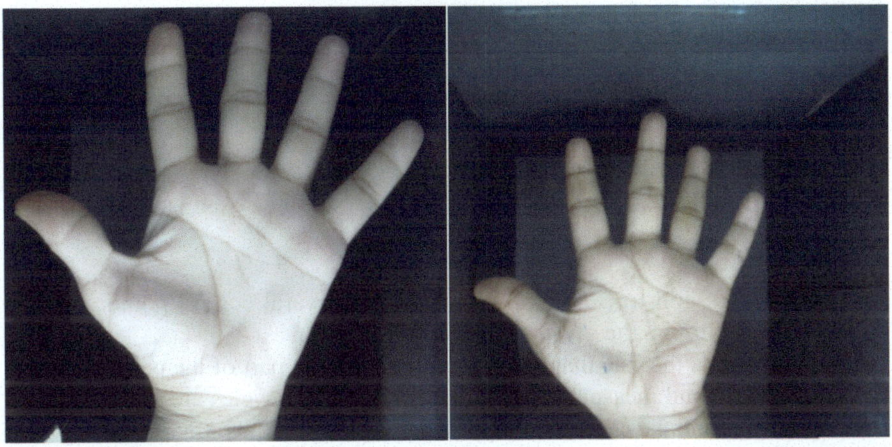

**Fig. 2** An example of the location of the fingers of the palm that are used in the proposed method

account the location of the fingers. The method that is described in this chapter requires the fingers to be placed in an open and relaxed palm laid on the surface of the scanning device. An example of the image of such a palm on Fig. 2 is shown.

Free palm images for all identifiable hands can also be used. In this case, all identifiable must have the same finger arrangement.

Initially, using a scanner, the palm image is converted into a digital file, which is downloaded to a computer system. The resulting color image of the palm is converted to a bitmap (Fig. 3).

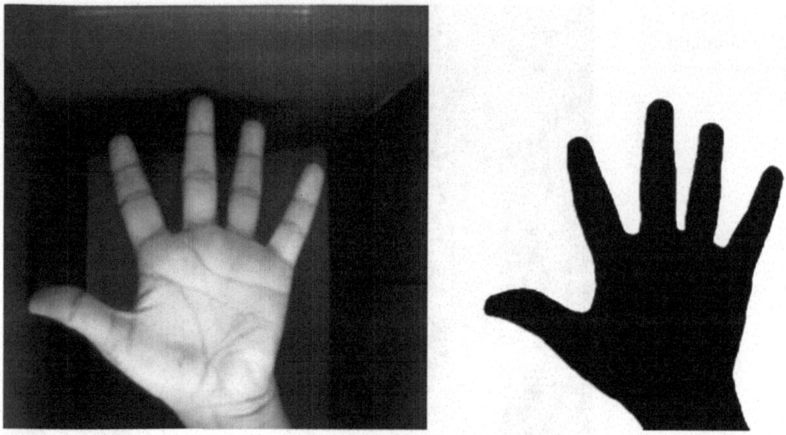

**Fig. 3** An example of converting a color image of the palm into a bit

For each color image, the program evaluates the brightness and color characteristics and, based on their analysis, selects the binarization threshold such that the inner palm region is black. For the image presented on Fig. 3, a binarization threshold of 85% is selected. It is also possible to invert colors in the image.

Biometric characteristic features are extracted using the entered additional geometric structures. These include:

- top rectangle;
- bottom rectangle;
- top line;
- main line;
- first bottom line;
- second bottom line.

Rectangles are constructed taking into account several control points, and the lines must pass through the given control points.

The main line is the line that runs through the widest part of the palm along the X axis. To search and measure the main line, the vertical cycles are organized, and inside it is a horizontal cycle, which bypasses the pixels of the bitmap, starting from the upper left corner. This cycle searches for a continuous horizontal line consisting of black pixels. Counter counts black pixels going one after another. If the black line is interrupted, then the horizontal cycle is interrupted, and the counter value is compared with the maximum recorded before, and when the maximum is exceeded, it is replaced. The program that implements the method was developed in the Delphi environment using the Object Pascal language. The listing fragment for the loop is described by the following instructions:

```
max := 0;
 For i := 0 to n - 1 do
  begin
   c := 0;
   For j := 0 to m - 1 do
    begin
     if (M1[j, i] = 0) and (c > 0) then
      Break
     else
      if M1[j, i] = 1 then
       begin
        if c = 0 then
         begin
          x := j;
          y := i;
         end;
        Inc(c);
       end;
    end;
   if c > max then
    begin
     max := c;
     Line1_1.X := x;
     Line1_1.Y := y;
     Line1_2.X := Line1_1.X + max - 1;
     Line1_2.Y := Line1_1.Y;
    end;
  end;
```

The top line is the line that runs through the widest part of the palm, starting from the top right after the fingers. To search and measure the top line, the X coordinate of the center line center is first calculated. Then, a vertical cycle is organized with the pixel traversal of the bitmap from the bottom and up starting from the coordinate Y of the main line, and an internal horizontal cycle is realized that implements the pixel traversal of the bitmap from the middle of the main line to the end of the image from left to right. In the process, the counter counts continuously moving black pixels along the X axis. If black pixels are interrupted, then the horizontal cycle is interrupted, and the counter value is compared with the maximum recorded before, and when the maximum is exceeded, it is replaced.

A fragment of the program that implements the formation of the upper line is represented by the following sequence of commands.

```
k := Line1_1.X + (Line1_2.X - Line1_1.X) div 2;
 Line2_1.X := k;
 max := 0;
 For i := Line1_1.Y downto 0 do
  begin
   c := 0;
   For j := k to m - 1 do
    begin
     if (M1[j, i] = 0) and (c > 0) then
      Break
     else
      if M1[j, i] = 1 then
       begin
        if c = 0 then
         y := i;
        Inc(c);
       end;
    end;
   if c > max then
    begin
     max := c;
     Line2_1.Y := y;
     Line2_2.X := Line2_1.X + max - 1;
     Line2_2.Y := Line2_1.Y;
    end;
  end;
```

After that, the coordinates of the upper line are calculated. To do this, a cycle along the X axis is performed, starting from the right edge of the area found in the previous step, to the left to the beginning of the palm image. In this case, a check is made for the meeting of the first white pixel that has come across. As soon as a white pixel is encountered, the X coordinate is fixed and the loop stops working. A listing fragment of a program that implements the described process consists of the following commands and operators.

```
For j := Line2_2.X downto 0 do
 if M1[j, Line2_2.Y] = 0 then
  begin
   Line2_1.X := j + 1;
   Break;
  end;
```

The next biometric parameter is the interval (distance) between the main and upper line. The interval is calculated as the difference between the coordinates along the Y axis of the main and upper lines.

```
Interval := Line1_1.Y - Line2_2.Y;
```

The first bottom line is formed as follows. The interval departs from the main line down the Y axis and searches for coordinates along the X axis at this point. To do this, execute a cycle along the X axis from left to right from the beginning to the end of the image. In this case, the X coordinates of the beginning and end of the first continuous sequence of black pixels encountered are recorded. As soon as the sequence of black pixels is interrupted, the cycle stops. A fragment of the listing that implements the formation of the first bottom line has the following structure.

```
Line3_1.X := -1;
  Line3_1.Y := Line1_1.Y + Interval;
  Line3_2.Y := Line3_1.Y;
  For j := 0 to m - 1 do
    if (M1[j, Line3_1.Y] = 1) and (Line3_1.X = -1) then
      Line3_1.X := j
    else
      if (M1[j, Line3_1.Y] = 0) and (Line3_1.X <> -1) then
        begin
          Line3_2.X := j - 1;
          Break;
        end;
```

The second bottom line is determined by moving the interval down along the Y axis and searching for coordinates along the X axis at this location. To do this, execute a cycle along the X axis from left to right from the beginning to the end of the image. At the same time, the X coordinates of the beginning and end of the first continuous sequence of black pixels are recorded. As soon as the line of black pixels is interrupted, the cycle stops working.

A fragment of the program that implements the definition and formation of the second lower line is as follows.

```
Line4_1.X := -1;
  Line4_1.Y := Line3_1.Y + Interval;
  Line4_2.Y := Line4_1.Y;
  For j := 0 to m - 1 do
    if (M1[j, Line4_1.Y] = 1) and (Line4_1.X = -1) then
      Line4_1.X := j
    else
      if (M1[j, Line4_1.Y] = 0) and (Line4_1.X <> -1) then
        begin
          Line4_2.X := j - 1;
          Break;
        end;
```

The main biometric characteristics of the palm that are used to implement the method are shown in Fig. 4.

To calculate the area, it is necessary to calculate the parameters of the rectangle areas.

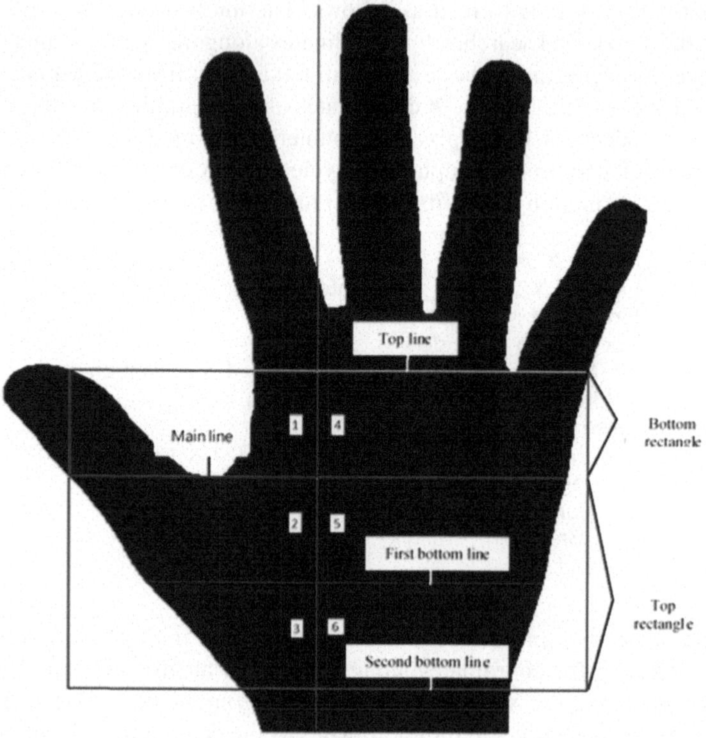


**Fig. 4** Display of the basic biometric characteristics on the palm image that implement the proposed method

1. The main rectangle is the height from the top line to the bottom line, the width from the beginning of the main line on the left to the end of the top line on the right.
2. Six areas. The main rectangle is divided along the Y axis in the place where the middle of the main line is located. Then, by means of the main line and the first lower horizontal line, the main rectangle is divided.
3. The upper rectangle is the upper part of the main rectangle, divided by the main line along the X axis.
4. The bottom rectangle is the bottom of the main rectangle, divided by the main line along the X axis.

After six areas are selected, the calculation of the areas covered by black pixels representing the corresponding part of the palm follows. Next, we calculate the relations that allow us to process images of the palms presented at different scales.

For sufficient identification, the following relations are used.

1. The ratio of the area occupied by black pixels in the main rectangle to the six areas of the rectangles included in the main rectangle, which are taken into account alternately.
2. The ratio of the area occupied by black pixels in the lower rectangle to the area in the upper rectangle.

3. The ratio of the length of the main line along the X axis (the widest part of the palm) to:
   (a) Top line;
   (b) First bottom line;
   (c) Second bottom line.

To implement the identification process, a database is created (electronic directory), which is represented by a file of a structural type. An entry is added to the directory that contains the name of the reference and all calculated relationships.

Identification is carried out as follows.

1. Checks for the presence of a directory file.
2. The directory file is opened.
3. In the directory file, the all entries are analyzed. In this case, the resulting array of structures is filled. Each structure contains.
   (a) Reference serial number in the directory file;
   (b) The name of the reference;
   (c) An array in which values 1 or 0 are placed.
   If the absolute value of the difference in the ratio of the reference and the corresponding ratio of the recognized object is less than the specified accuracy (confidence interval), then 1 is written to the array, otherwise 0. Recording to the array of results is performed at the position that corresponds to the position in the array of relations.
   (d) Sum of ones (matches) recorded in the previous paragraph.
4. Sorting by the sum of ones (matches) of the resulting array of structures in descending order.
5. Output the resulting array to a table on the main form.

Thus, in the table of recognition results on the main form, the references with which there is the greatest coincidence are shown first. Next in descending order are the references with fewer coincidences.

## 3 Experiment Description

Based on the described method, computer simulation was performed using the Delphi environment and Object Pascal language. Fragments of the program that describe the algorithms for determining the selected biometric characteristics are presented in the previous section. The interface of the program that implements the model of the method on Fig. 5 is presented. The interface is divided into several fields.

- Field displaying palm shape and basic biometric characteristics.
- Field displaying quantitative results of calculating biometric characteristic features for a selected palm image.
- Field for displaying identification results.

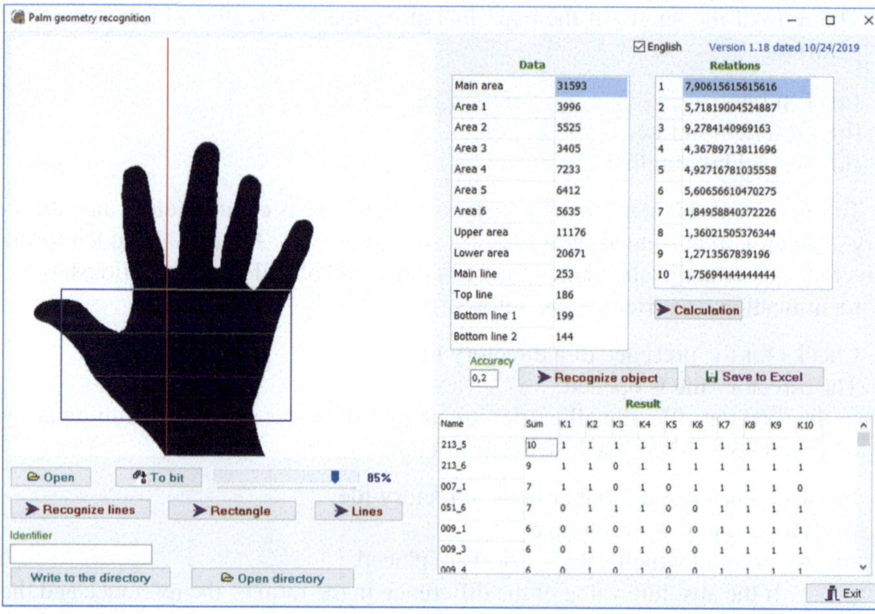


**Fig. 5** The interface of the program that implements the method

For the experiment, palm images obtained from a scanning device and having clear palm borders on the rest of the background were used [2]. The resulting image was binarized for different binarization thresholds. The obtained quantitative values of the relations of biometric characteristics are recorded in the database and they are assigned the identifier of the person to whom this palm belongs. The database was formed for images of the palms of the right and left hands of a person.

The identification process is carried out by downloading an image from a scanning device and determining the necessary quantitative values of the ratios of the selected biometric characteristics. After loading the calculated relationship values, the nearest reference value recorded earlier in the database is searched.

At the initial time, the nearest references are searched for by the value of the main line or the upper line. From the selected nearest references, the remaining biometric parameters are selected and compared with identical parameters from the image at the system input. If the number of matched quantitative values exceeds the set threshold, then the identifier of the nearest reference identifies the identified person.

Part of the biometric characteristics of the shape of the palm may coincide for the impostor. However, most are mostly different for different people. In the process of recognition, comparisons are made with the references of each group in turn. The percentage of matching values for each selected accuracy value is determined. If the calculated percentage exceeds the selected threshold value, then the corresponding identifier identifies the person. An example of identifying of "own" on Fig. 6 is presented.

In this example, five reference values for one identifiable are present in the database. The comparison was carried out for accuracy from 0.1 to 0.5. Here, the second reference is represented by an unsuccessful hand position on the scanning device. With an accuracy of 0.1 and 0.2, not all references have a complete match between the calculated ratios. An accuracy of 0.3 gives a complete match of all ratio values for the four references of this user.

The results of comparisons of the formed relations with the reference values belonging to another person at intervals of 0.1, 0.2, and 0.3 give significant deviations. Approaching the full value appears at intervals of 0.4 and above (Fig. 7).

In this Fig. 7 the last two groups of values identify their own, and the remaining values identify the impostor.

These thresholds are selected for the scanner used. There is also a need for a specific palm position. The palm should be fully displayed on the generated image.

Authentication can be done by analyzing the percentage of matches for each user.

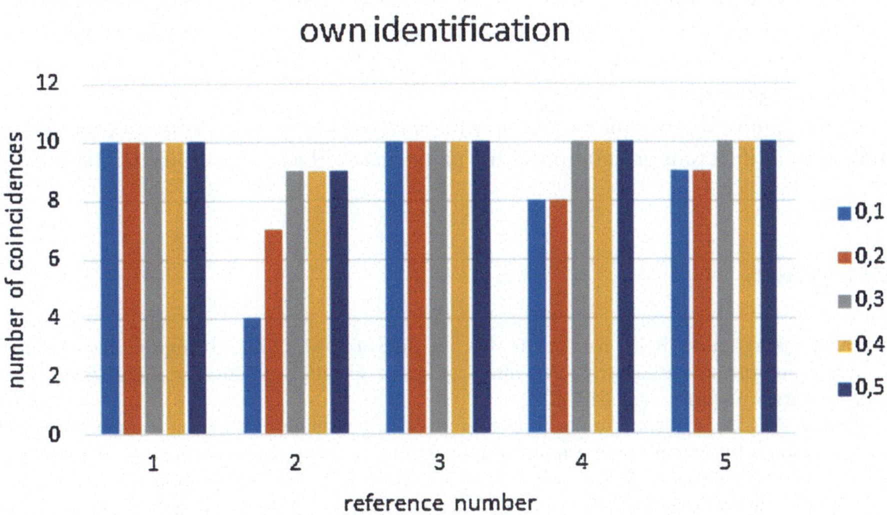


**Fig. 6** An example of identification for "own"

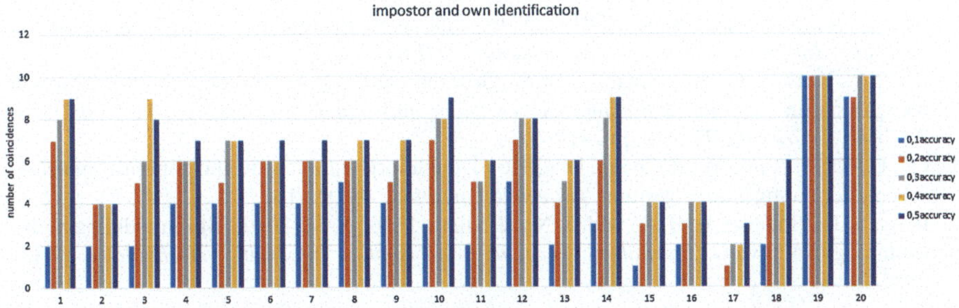


**Fig. 7** An example of the identification of one's own and impostor

## 4 Conclusion

The palm on the image has large variations in parameters, although the method is considered simple to implement. Identification by palm image depends on location on the scanning surface and on the scanner itself. The ratio of the geometric shapes of the palm on the scanned image is important.

The paper considers the method of biometric identification of a person by the image of his inner side of the palm. The method made it possible to simplify the identification process by selecting biometric characteristics belonging to the palm region, which remains practically unchanged. The method practically does not take into account the geometric shapes of the fingers and their location. In this case, there are certain requirements for the location of the palm on the scanning device. The confidence interval for the images that participated in the experiment was determined. The experiment showed that the more standards used in the database to identify a single user, the higher the accuracy of identification. When using ten or more reference values, the permissible FFR and FAR values for this identification method are determined.

In the future, it is planned to modify the method for the possibility of processing images with different orientation of the palms, as well as use various databases.

## References

1. Hennings-Yeomans, P. H., Kumar, B. V. K., & Savvides, M. (2007). Palmprint classification using multiple advanced correlation filters and palm-specific segmentation. *IEEE Trans. Info Forensics & Security, 2*(3), 613–622.
2. Kumar, A. (2008). Incorporating cohort information for reliable Palmprint authentication. In *Sixth Indian Conference on Computer Vision, Graphics & Image Processing* (pp. 583–590).
3. Genovese, A., Piuri, V., & Scotti, F. (2014). *Touchless Palmprint recognition systems*. Berlin: Springer International Publishing.
4. Qiu, Z., et al. (2019). Local discriminative direction extraction for Palmprint recognition. In *Biometric recognition. 14th Chinese conference, CCBR 2019, Zhuzhou, China, October 12–13, 2019, proceedings* (pp. 3–11).
5. Zhang, D., Kong, W. K., You, J., & Wong, M. (2003). On-line palmprint identification. In *IEEE trans. Patt. Anal. Machine Intell* (Vol. 25, pp. 1041–1050).
6. Sneha, M., & Dhananjay.M. (2013). Palmprint authentication using SIFT. *International Journal of Engineering Research & Technology (IJERT)., 2*(9), 2439–2444.
7. Bilal, A., Youssef, C., & Amina, S. (2018). Geometrical local image descriptors for palmprint recognition. In *International conference on image and signal processing 2018 (ICISP 2018), Jul 2018, Cherbourg, France* (pp. 419–426).
8. Fei, L., Xu, Y., Tang, W., & Zhang, D. (2016). Double-orientation code and nonlinear matching scheme for palmprint recognition. *Pattern Recognition., 49*, 89–101.
9. Inass, S. H., Shamsu, L. B. S., Md, J. N., & Nilam, N. B. A. S. (2019). Multimodal palmprint technology: A review. *Journal of Theoretical and Applied Information Technology, 97*(11), 2882–2896.

# Search of Informative Biometric Characteristic Features of the Palm Based on Parallel Shift Technology

**Sergey Yuzhakov, Sergii Bilan, Stepan Bilan, and Mykola Bilan**

## 1 Introduction

Modern technologies in the field of biometric personality identification are developing rapidly. For biometric identification of a person, a large number of biometric characteristics are used, which are divided into static and dynamic [1–3]. These characteristics constantly belong to one person throughout life and can change with his age. In this case, the static biometric characteristics (fingerprint, face image, geometric shape of the palm, ear, iris and retina of eye, etc.) can be faked without much difficulty. Two-dimensional characteristics can be photographed, and three-dimensional can be copied and recreated using a 3D printer. Dynamic biometric characteristics (voice, gait, handwriting and keyboard handwriting, etc.) can be faked with great difficulty, therefore they are more reliable.

Improving the accuracy and reliability of identification is achieved through the use of several biometric parameters, both static and dynamic [1–3]. This is due to the fact that the quantitative values of some biometric characteristics for different people may coincide. This is especially true for relatives. Modern scanning devices capture biometric characteristics with high quality [4]. Many of them include

S. Yuzhakov (✉)
Haisyn Department of the SFS General Directorate in Vinnytsia Region, Vinnytsia, Ukraine
e-mail: yserg74@meta.ua

S. Bilan
Onseo Company, Vinnytsia, Ukraine

S. Bilan
State University of Infrastructure and Technology, Kyiv, Ukraine
e-mail: bilan_sm@gsuite.duit.edu.ua

M. Bilan
The Municipal Educational Institution Mayakskaya Secondary School, Mayak, Moldova
e-mail: bilannn@mail.md



S. Bilan et al. (eds.), *Biometric Identification Technologies Based on Modern Data Mining Methods*, https://doi.org/10.1007/978-3-030-48378-4_10

pre-processing functions for clearly extracting the required biometric parameter and removing uninformative information and noise.

Almost all biometric characteristics are a combination of a large number of distinctive elements [4]. A fingerprint consists of a large number of lines of various shapes. Moreover, for different patterns of fingerprints, individual lines can have the same shape. When identifying by faces of different people, lip shapes, forehead shape, eye color, etc. may coincide separately. However, faces are identified as for different people.

One of the commonly used biometric characteristics is the geometric shape of the palm of a person's hand [5–13]. Special scanning devices that require a rigid hand position on the scanning surface read this biometric characteristic. To date, more than 90 biometric characteristics are known that distinguish two palms belonging to different people. However, part of the biometric characteristics of the palm may coincide for different people. Therefore, the analysis of all biometric characteristics of the palm gives the greatest accuracy of identification. Moreover, the analysis of each biometric characteristic of the palm requires separate algorithms and tools that complicate the system as a whole.

In this regard, the task is to determine the biometric characteristics of the palm, which determine the most significant differences for all the palms at the entrance of system. Since the analysis of all the biometric characteristics of the palm requires a lot of time and effort in the work, the biometric characteristics described by Bilan M. and Bilan A. are analyzed. According to these characteristics, the geometric shape of the palm is described by the areas of individual sections of the palm image.

To solve the problem of searching for informative biometric characteristics, parallel shift technology is used (PST) [14–18], which allows to analyze the image area of any shape. PST gives users the opportunity to uniquely evaluate the geometric structure of the image.

## 2 Parallel Shift Technology

Sergey Yuzhakov and Stepan Bilan proposed the parallel shift technology [14–18] in the early 2000s. This technology is mainly aimed at analyzing the geometric shapes of images using one parameter—area. Area in PST is defined as a collection of pixels belonging to an image [14, 15]. This area can be defined as a collection of pixels within a certain closed border, as well as a set of different groups of pixels scattered throughout the image field [15].

PST is that for the analyzed image, a copy is created that moves in different directions [14]. At each time step of the shift, the intersection area of the original image and its copy is determined (Fig. 1).

During the shift time of the image copy, the function of the area of intersection (FAI($\varphi$)) in the shift direction $\varphi^0$ is formed. The FAI defines a set of quantitative values that correspond to the area of intersection at each time shift step. FAIs are

**Fig. 1** Formation of the area of intersection (highlighted in green) of the initial image and its copy shifted by 120 time steps in the direction of shift by $0^0$

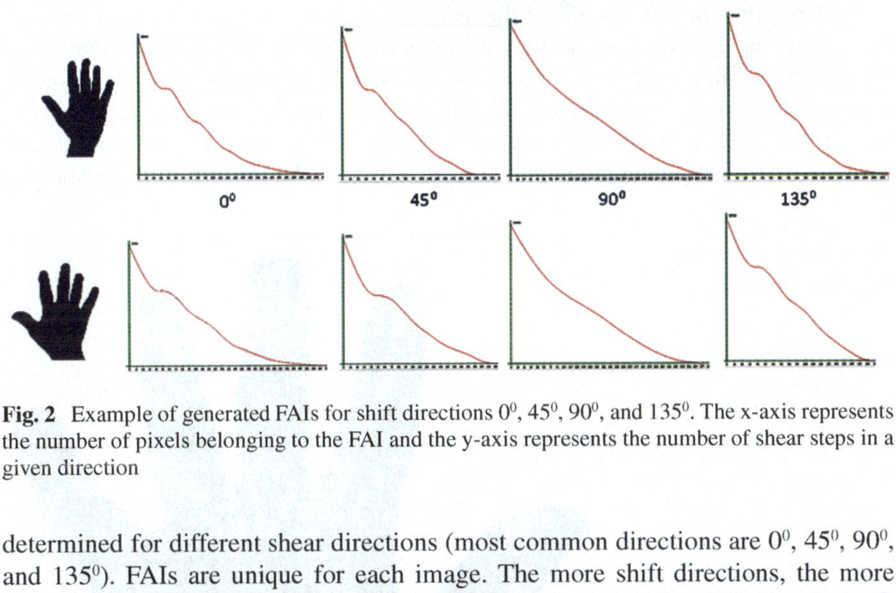


**Fig. 2** Example of generated FAIs for shift directions $0^0$, $45^0$, $90^0$, and $135^0$. The x-axis represents the number of pixels belonging to the FAI and the y-axis represents the number of shear steps in a given direction

determined for different shear directions (most common directions are $0^0$, $45^0$, $90^0$, and $135^0$). FAIs are unique for each image. The more shift directions, the more accurate the image description. An example of formed FAIs for shift directions $0^0$, $45^0$, $90^0$, and $135^0$ on Fig. 2 is shown.

For the same image, the FAIs in the opposite directions are the same. For example, FAI($0^0$) = FAI($180^0$). Therefore, the shift directions from $0^0$ to $179^0$ are used.

The advantage of PST is the use of a single geometric parameter, which is easily determined using computer technology. There are also many solutions for improving image analysis using this technology by introducing additional coefficients and standard functions of the area of intersection.

## 3 Analysis of the Geometric Shape of the Palm

To analyze the geometric shape of the palm, the palm image sections were used, which are determined by the method proposed by Bilan M., Bilan A., and Bilan S. This technique does not take into account parts of the palm that can change their location on the scanning device. For example, fingers may have a different arrangement. Only parts of the palm that are not changed are taken into account. These parameters include the parts that make up the middle part of the palm, with a large area.

The technique consists in the fact that the binary image of the palm is divided into six areas of the whole middle part of the palm, taking into account the location of the thumb (Fig. 3).

Rectangles are constructed taking into account control points located on the image of a person's palm. The upper side of the rectangle is determined by the line that runs through the widest part of the palm immediately after the end of the lower part of the image of the fingers. Typically, this line touches the hollow between the fourth and fifth fingers on the left for the image of the left palm. Therefore, the location of the fingers does not affect the location of the top line of the rectangle.

To determine the bottom side of the rectangle, a line is drawn that passes through the widest, inextricable part of the palm. Usually, this line is tangent to the cavity between the thumb and forefinger and is parallel to the horizontal axis. The distance between these lines is calculated. To construct the bottom side of the rectangle, these calculated three distances between the top line and the second line passing through the widest part of the palm recede down from the top side of the rectangle.

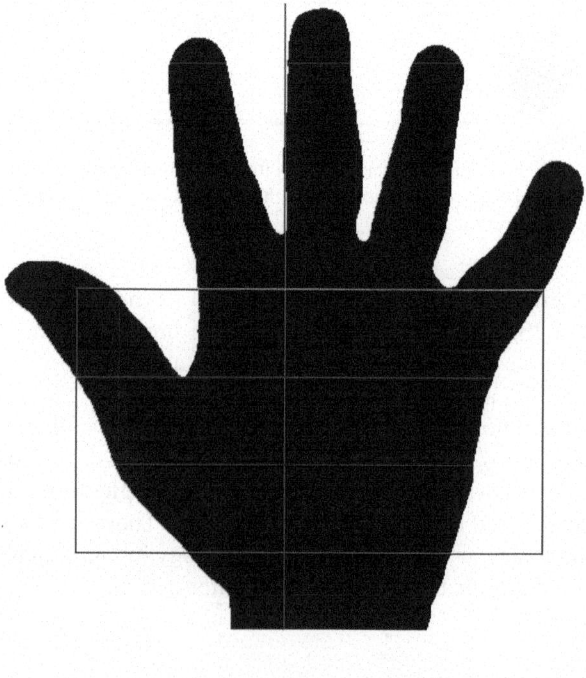

**Fig. 3** A binary image of the palm, divided into six areas

In Fig. 3 is the second line from the top. At each such distance, a horizontal line is formed, dissecting the rectangle into several rectangular areas.

The right side of the rectangle is formed vertically down from the point of intersection of the top line with the right edge of the palm to the intersection of the bottom horizontal line of the rectangle. The left side is formed by a vertical line down from the point of intersection of the upper horizontal line with the extreme point of the thumb of the palm.

Two horizontal lines are drawn inside the formed rectangle, which are located at a distance equal to the distance between the two upper lines, which are described above. Thus, the rectangle is divided into three equal areas. Then a vertical line that is perpendicular to the center point of the line of the widest part of the palm is drawn. To draw a vertical line, the center point of the line of pixels constituting the widest part of the palm is searched (the line containing the largest inextricable number of pixels belonging to the palm image).

A vertical line divides the resulting three regions of the rectangle into six regions. The rectangular image of each such area contains pixels belonging to the palm image and background. Black pixels belonging to the palm make up a specific geometric shape or several areas.

Thus, as a result of the transformations, six images consisting of white and black pixels are obtained. Black pixels are part of the human palm image. Each image for one palm has slight differences for similar images of the other palm (Fig. 4).

PST is used to analyze the presented images. For each of the six images of one palm, FAIs are generated for the shift directions $0^0$, $45^0$, $90^0$, and $135^0$. To construct these functions of intersection of areas, a different number of nonzero shifts is used (the time step at which FAI $\neq$ 0). Examples of graphic images of the FAI for six images obtained from the image of one palm on Fig. 5 are presented.

**Fig. 4** Examples of six formed images for one palm

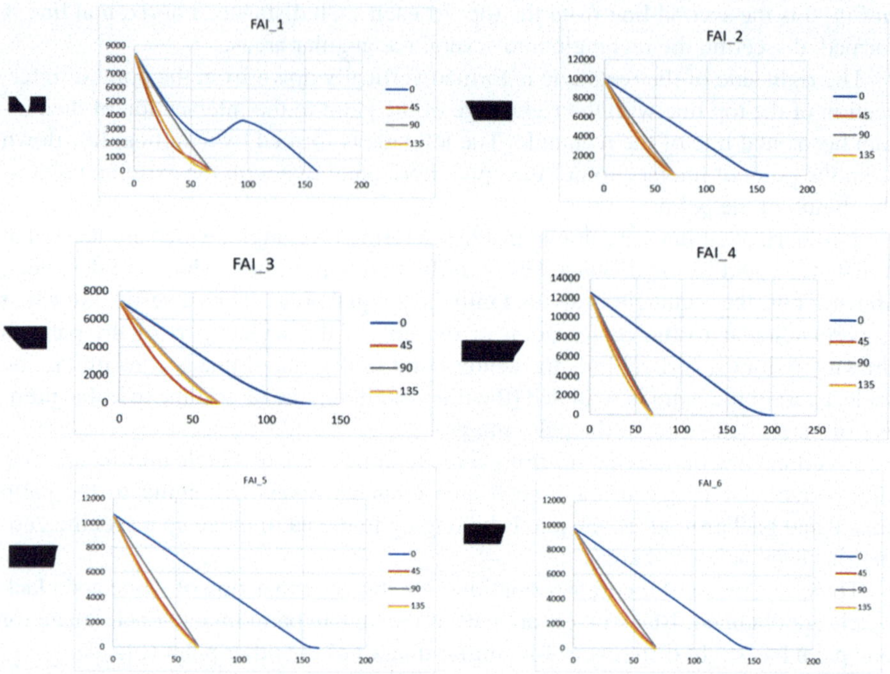


**Fig. 5** FAI examples for six images from one palm

A reference database is formed, which consists of ten FAIs for each shift direction ($0^0$, $45^0$, $90^0$, and $135^0$) and for one identifier that identifies one person. Thus, for each person ten standards of images of one palm are formed in the form of quantitative values of the functions of the area of intersection. Standards are formed by scanning one palm ten times. Moreover, scanning can be performed at different times during the formation of the database. The more reference values recorded in the database, the more accurate the identification.

For each person, reference values are generated for each reference FAI, and relationships between average FAI values for one of the six palm images are also determined. An array of FAI relationships is generated, represented by Table 1. Each column of Table 1 indicates the ratio of the average FAI of one direction to average FAIs generated for other shear directions. This takes into account the circumstances that the number of values forming the FAI is different. Therefore, each average FAI value is calculated differently using different arrays of numbers.

Thus, the reference database for each person contains ten standards, each of which contains the FAI of each of the six palm images. Moreover, the FAI is presented for the four directions of the shift since they are easily implemented on the orthogonal covering of the array ($0^0$, $45^0$, $90^0$, and $135^0$). Also, for each FAI, the average value is determined in the database and for each of the six images, an array of relations of average FAI values is formed according to Table 1.

When the database is formed, the process of identification and authentication begins. A binary image of a person's palm from a scanning device is fed to the input

**Table 1** The formation of an array of relations

| Relationship for FAI($0^0$) | Relationship for FAI($45^0$) | Relationship for FAI($90^0$) |
|---|---|---|
| $\frac{FAI(0^0)}{FAI(45^0)}$ | $\frac{FAI(45^0)}{FAI(90^0)}$ | $\frac{FAI(90^0)}{FAI(135^0)}$ |
| $\frac{FAI(0^0)}{FAI(90^0)}$ | $\frac{FAI(45^0)}{FAI(135^0)}$ | |
| $\frac{FAI(0^0)}{FAI(135^0)}$ | | |

of the system. Six images are extracted from the binary palm image for which the FAI is determined in four shift directions ($0^0$, $45^0$, $90^0$, and $135^0$). For each FAI, average values are determined and for each image an array of relations of average FAI values is formed according to Table 1.

The resulting array of relations is compared with the reference arrays and coincidence arrays are formed. Each array of coincidences has the same dimensions as the array of relations i.e. consists of six numbers. An array of coincidences contains only 1 or 0 at each position. If the value of the array of relations differs from the value in the reference array of relations less than the selected value D, then 1 is generated in the array of coincidences. Otherwise, 0 is generated.

If, for each reference group of relations of one user, the number of coincidences is greater than the specified value, then the identifier of the corresponding standard identifies the person at the input. If the percentage of the number of coincidences is less than the selected threshold value, then the person at the input is not identified, since his reference values are not in the database.

The use of PST for the image of the entire palm does not provide the necessary identification result. This is due to the fact that the palm of a person's hand may not have a complete image or vice versa may have additional elements (for example, a part of the hand is present along with the palm). Also, the location of the fingers on the scanning device gives different forms of FAI obtained in the perpendicular direction of shift relative to the fingers.

If you use the upper cut off part of the palm, consisting of four fingers (Fig. 6), it is difficult to determine who owns this palm. FAIs obtained for the upper part of one palm differ for the same palm with different finger positions. If the gaps between the fingers are greater, then the distances between the upper and lower extrema of the FAI are also greater for the directions of movement $0^0$, $45^0$, and $135^0$. In practice, the shape of the FAI does not change if the shift direction coincides with the direction along the fingers of palm image (in the example shown in Fig. 6—it is $90^0$ direction).

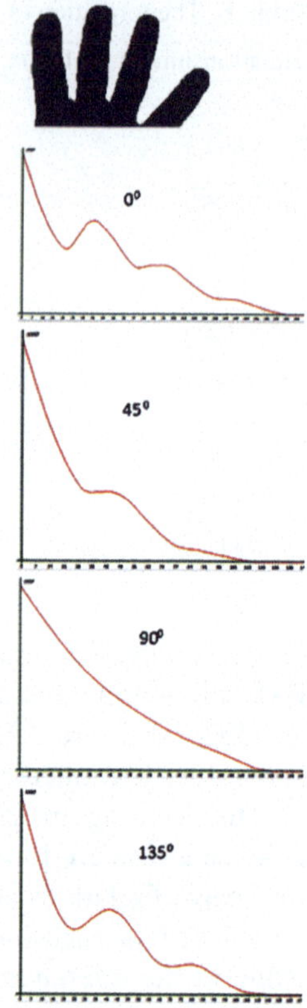


**Fig. 6** Examples of upper palm FAIs obtained in directions $0^0$, $45^0$, $90^0$, and $135^0$

In this regard, PST is not used for images containing the fingers of the palm. These parts of the images did not participate in the experiment. PST is successfully used to evaluate the geometric shape of each finger individually. However, for this it is necessary to use additional methods for selecting images of the fingers of the palm, taking into account their orientation.

## 4 Experiment and Basic Acceptable Values

For carrying out the experiment, IIT Delhi Palmprint Image Database [5] palm image database was used, the use of which the authors received permission. Images were selected with a full palm image. The method is advisable if the palm is fully located in the scanning field. Left palms and mirror reflections of the right palms were used. For all palms used, their binary images were obtained after preliminary processing.

Threshold image processing does not always give a clear highlighted palm image. Therefore, additional image processing methods were used. These methods are based on the use of the theory of cellular automata, which made it possible to remove small groups of black pixels that do not belong to the palm image [19–21]. In cellular automata, the AND function was used, which allows you to remove small isolated groups of pixels in a certain number of time steps.

To remove individual groups of black pixels, it was also carried out by implementing the Radon transform and constructing projections in six directions [22–24]. The projections obtained were analyzed, as a result of which groups of black pixels not belonging to the palm image were determined. In the corresponding projections of the Radon transform, the directions and lines on which the noise pixels are located were determined. Along these lines, black pixels switched to white pixels.

The database used does not contain more than ten images of the same palm. Therefore, a smaller number of palm images was used to create the pattern database. Moreover, the experiment showed the reliability of the identification results.

As a result of the experiment, palm images from the database were input to the system and arrays of relations and matches for each identified palm were formed. The generated relationship arrays for each of the six images were compared with the reference relationship arrays and coincidence arrays were formed for different confidence intervals. The 0.01; 0.02; 0.03; 0.04; 0.05; 0.06; and 0.07 confidence intervals were used.

The analysis of the obtained coincidence arrays by two characteristics was carried out.

The first characteristic determined the percentage of matches for the "impostor" and the percentage of matches for "own." The difference between the obtained values for each of the six images was also determined. As a result of the experiment, dependency graphs were obtained for coincidence arrays with confidence intervals of 0.04 and 0.05 (Fig. 7).

The obtained graphs show that the most informative are 1, 2, 5, and 6 images extracted from the binary image of the palm. These images give the largest percentage in identifying "own" and the smallest percentage of matches in identifying "impostor." For 1 and 2 images, the difference value exceeds the percentage value of matches to identify the "impostor."

The second characteristic was determined as follows.

For each array of matches, the number of matches was determined (number of 1 was counted). If the number of matches exceeded the specified number M (in the experiment, M = 4 and M = 5 were used), then for this array, 1 is formed. For each M, the percentage of coincidences was determined to identify "own" and "impostor." As a result of the experiment, we obtained the dependencies of the percentage of coincidence for confidence intervals 0.04 and 0.05 (Fig. 8).

An analysis of the obtained experimental results showed that the fourth image is the least informative image, which coincides with the previous version based on the first characteristic.

In accordance with both characteristics used, the first, second, and sixth images are the most informative. These images give a very low percentage of matches when identifying an impostor. Although the fourth and fifth images give 100% matches when identifying "own," there is a high percentage of matches when identifying an "impostor." For such images, it is necessary to increase the threshold for the percentage of coincidences in order to reduce the FAR value to 0 and also significantly reduce the FRR value.

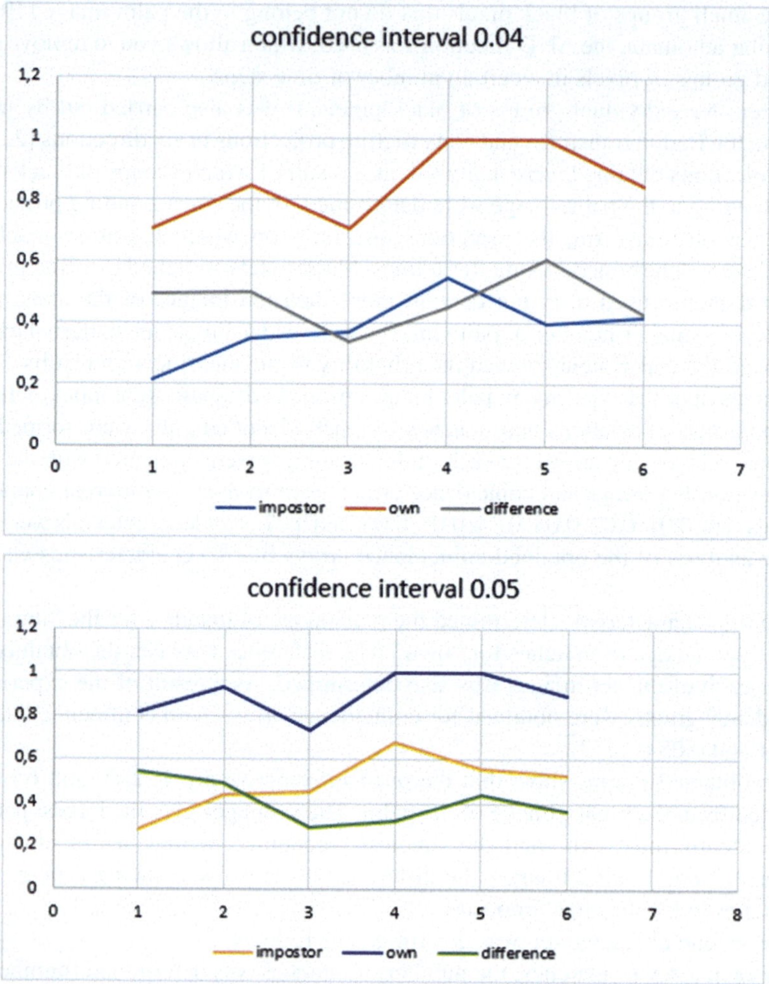


**Fig. 7** Graphs of dependencies of percent matches for the "impostor" and "own," as well as the difference between them

The best results were shown with a confidence interval of 0.04. However, a confidence interval of 0.05 also gives a high result and can be used to identification.

## 5 Conclusion

For biometric identification, the geometric shape of the palm is often used. To analyze the geometric shape of the palm, simple scanning devices are used. However, among a large number of geometric characteristics, not all of them have a high

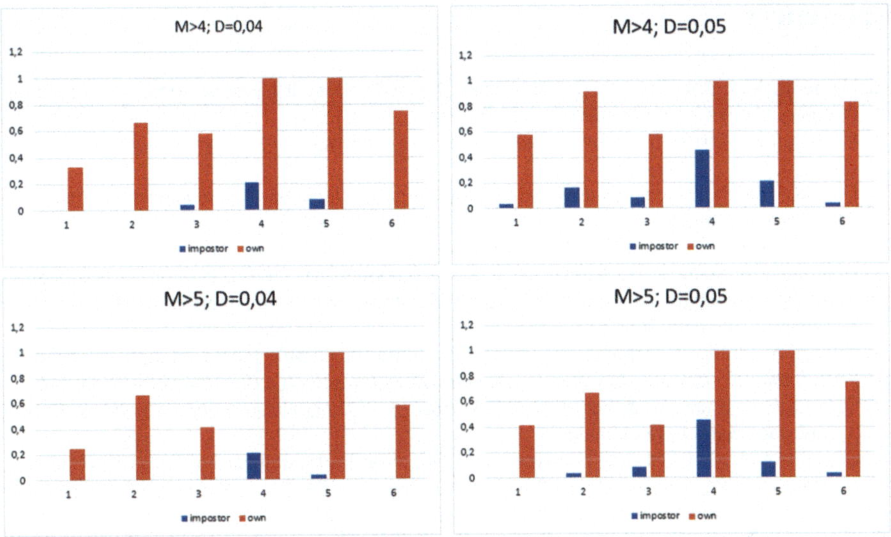


**Fig. 8** Graphs of the percentage of coincidence in coincidence arrays with a threshold value of M > 4 and M > 5 for confidence intervals 0.04 and 0.05

information content. This chapter solves the problem of finding the most informative parts of the palm image. With the help of parallel shift technology, information images of palm parts were determined, which made it possible to achieve high accuracy of biometric identification. A technique is proposed for analyzing the characteristics of images of parts of the palm on the basis of two characteristics that made it possible to select the necessary threshold and determine the optimal confidence interval (0.04 and 0.05), which gives the greatest identification accuracy for the used method. Parallel shift technology operates with one parameter—area. Analysis and determination of the image area does not require complex computing resources. Parallel shift technology can be used to analyze images of the entire palm. However, a different arrangement of the fingers of the palm gives different functions of crossing the areas, which can lead to false identification. Therefore, it is better to use images of such parts of the palm that do not change their shape in any location. These parts include fingers, each individually. PST does not give high results when evaluating patterns of skin folds in the palm image.

In future studies, it is planned to analyze other biometric characteristics of the palm to determine the most informative of them. In future studies, it is planned to combine parallel shift technology and image identification theory based on the Radon transform, since both of these directions form characteristic features by analyzing images in different directions of shift. This allows a comprehensive analysis of the geometric shape of the image.

## References


1. Nambiar, A., Bernardino, A., & Nascimento, J. C. (2019). Gait-based person re-identification: A survey. *ACM Computing Surveys, 52*(2), 33.
2. Fairhurst, M. (2019). *Biometrics: A very short introduction (very short introductions)*. Oxford: Oxford University Press.
3. Das, R. (2018). *The science of biometrics: Security Technology for Identity Verification*. Abingdon: Routledge.
4. Li, S. Z., & Jain, A. K. (2015). *Encyclopedia of biometrics*. Berlin: Springer.
5. Hennings-Yeomans, P. H., Kumar, B. V. K., & Savvides, M. (2007). Palmprint classification using multiple advanced correlation filters and palm-specific segmentation. *IEEE Trans. Info Forensics & Security, 2*(3), 613–622.
6. Kumar, A. (2008). Incorporating cohort information for reliable Palmprint authentication. In *Sixth Indian Conference on Computer Vision, Graphics & Image Processing* (pp. 583–590).
7. Genovese, A., Piuri, V., & Scotti, F. (2014). *Touchless Palmprint recognition systems*. Berlin: Springer International Publishing.
8. Qiu, Z., et al. (2019). Local Discriminative Direction Extraction for Palmprint Recognition. In *Biometric Recognition. 14th Chinese Conference, CCBR 2019, Zhuzhou, China, October 12–13, 2019, Proceedings* (pp. 3–11).
9. Zhang, D., Kong, W. K., You, J., & Wong, M. (2003). On-line palmprint identification. *IEEE Trans. Patt. Anal. Machine Intell., 25*, 1041–1050.
10. Sneha, M., & Dhananjay.M. (2013). Palmprint authentication using SIFT. *International Journal of Engineering Research & Technology (IJERT)., 2*(9), 2439–2444.
11. Bilal, A., Youssef, C., & Amina, S. (2018). Geometrical local image descriptors for palmprint recognition. In *International Conference on Image and Signal Processing 2018 (ICISP 2018), Jul 2018, Cherbourg, France* (pp. 419–426).
12. Fei, L., Xu, Y., Tang, W., & Zhang, D. (2016). Double-orientation code and nonlinear matching scheme for palmprint recognition. *Pattern Recognition., 49*, 89–101.
13. Inass S. H., ShamsuL B. S., Md Jan N., & Nilam N. B. A. S. (2019). Multimodal palmprint technology: A review. Journal of Theoretical and Applied Information Technology. 97(11): 2882–2896.
14. Bilan, S., & Yuzhakov, S. (2018). *Image processing and pattern recognition based on parallel shift technology*. Taylor & Francis Group: CRC Press.
15. Bilan, S., Al-zoubi, S. I., Yuzhakov, S., & Bilan, M. (2018). Description and recognition of symmetrical and freely oriented images based on parallel shift technology. In *5th International Conference on Mathematics and Computers in Sciences and Industry (MCSI)* (pp. 86–91). Athens: IEEE.
16. Bilan, S., Yuzhakov, S., & Bilan, S. (2014). Saving of etalons in image processing systems based on the parallel shift technology. *Advances in Image and Video Processing, 2*(6), 36–41.
17. Belan, S., & Yuzhakov, S. (2013). A homogenous parameter set for image recognition based on area. *Computer and Information Science, 6*(2), 93–102.
18. Belan, S., & Yuzhakov, S. (2013). Machine vision system based on the parallel shift technology and multiple image analysis. *Computer and Information Science, 6*(4), 115–124.
19. Wolfram, S. (2002). *A new kind of science*. Champaign, IL: Wolfram Media.
20. Bilan, S. (2017). *Formation methods, models, and hardware implementation of pseudorandom number generators: Emerging research and opportunities*. Philadelphia: IGI Global.
21. Bilan, S. M., & Al-Zoubi, S. I. (2019). *Handbook of research on intelligent data processing and information security systems* (p. 470). Philadelphia: IGI Global.
22. Bilan, S., Motornyuk, R., & Bilan, S. (2014). Method of hardware selection of characteristic features based on radon transformation and not sensitive to rotation, shifting and scale of the input images. *Advances in Image and Video Processing, 2*(4), 12–23.
23. Belan, S. N., & Motornyuk, R. L. (2013). Extraction of characteristic features of images with the help of the radon transform and its hardware implementation in terms of cellular automata. *Cybernetics and Systems Analysis., 49*(1), 7–14.
24. Motornyuk, R.L. (2013). Computer-aided methods for identifying images of moving objects based on cellular automata with a hexagonal coating. *Dissertation for the degree of candidate of technical sciences (UDC 004.932: 519.713 (043.3))*, Kiev: SUIT.

# Multitrait Selfie: Low-Cost Multimodal Smartphone User Authentication

**Zahid Akhtar, and Attaullah Buriro**

## 1 Introduction

Nowadays, smartphones have become pervasive personal computing platform. In fact, they have become personal assistants, which are widely being used not only for basic communications but also as a tool to store e-mail, personal photos, Internet banking and payments, and social networking. Thus, one of the biggest concerns is security and privacy leakage. Explicit authentication methods, e.g., PIN, password, graphical pattern, are the most common security strategy available on commercial smartphones [1]. They are susceptible to guessing or smudge attacks, besides being user-unfriendly and time-consuming [2]. Biometric-based user authentication on smartphones has been recently adopted. For instance, iPhone 6 s, Android KitKat mobile OS, and Fujitsu's Arrows NX F-04G employ fingerprint, face, and iris to recognize individuals, respectively. Nevertheless, they still face unresolved security[1] and usability issues [4, 5]. The average smartphone user checks their device 150 times per day.[2] If explicit passcode/biometric-based authentication takes 2 s, the typical user spends 5 min unlocking their device/app every day. In fact, recent studies have shown that about 34%, 47%, and 40% users did not use passcode, fingerprint, or any kind of authentication mechanisms, respectively, on their mobile

[1] Just 2 days after the iPhone5s hit the market, it was fooled by a fingerprint spoof [3].

[2] http://abcnews.go.com/blogs/technology/2013/05/cellphone-users-check-phones-150xday-andother-internet-fun-facts/.

Z. Akhtar (✉)
University of Memphis, Memphis, TN, USA
e-mail: zmomin@memphis.edu

A. Buriro (✉)
University of Bolzano, Bolzano, Italy
e-mail: aburiro@unibz.it



S. Bilan et al. (eds.), *Biometric Identification Technologies Based on Modern Data Mining Methods*, https://doi.org/10.1007/978-3-030-48378-4_11

devices citing low usability [4, 6]. Moreover, fingerprint scanner in smartphones most likely fails when it encounters dry fingers [7].

Researchers have lately proposed behavioral biometrics-based solutions for smartphone user authentication using built-in sensors such as accelerometer, gyroscope, and microphones. Some advantages of these techniques are minimal user interaction, unobtrusive data collection, and no additional hardware required [8]. For instance, the authors in [9] proposed a mobile framework *Sensec* based on sequence of actions that collects accelerometer, orientation sensor, touch screen, and magnetometer data for user authentication. While, Buriro et al. [10] employed a sensor enhanced touchstroke mechanism for smartphone biometric system, which analyzes how a person holds her phone and how she types her 4-digit free-text PIN. A specific five-finger touch gestures-based user recognition system for Apple iPad was presented in [11]. However, this method is not feasible for small touchscreens of typical smartphones. Blanco-Gonzalo et al. [12] studied handwritten signature recognition on mobile device.

A systematic analysis of prior methods and techniques for biometric smartphone user authentication show that they principally either require high user cooperation, work under constrained environment and protocol, use single modality that can be easily spoofed and mimicked (e.g., signature) [3] or do not yet exhibit low enough error rates. Moreover, no multimodal database using built-in three-dimensional sensors (e.g., accelerometer, gravity sensor, and magnetometer) is publicly available. In other words, it is well documented that unimodal biometric systems are unable to provide a high accuracy and security performances mainly due to noise in the sensed data, intra-class variations, inter-class similarities, and attacks [13]. Several above-mentioned problems can be solved or at least their impact reduced by fusing several biometric information sources; any such system is known as multibiometric systems [14]. Multibiometric systems offer many advantages over unimodal systems, such as significant improvement in the overall accuracy, mitigation of the effect of noisy input data, population coverage larger than the unimodal system [15, 16], greater resistance to spoofing such that they can be more robust than each corresponding unimodal system, even in the case when all biometric traits are spoofed [17]. Despite several advantages the state-of-the-art in unconstrained and unobtrusive multimodal smartphone biometric authentication is relatively underexplored field.

Thus, in this chapter, we aim to show that multimodal biometrics can be incorporated with smartphones in a user-friendly fashion and can also improve security and robustness. Specifically, this chapter presents an unconstrained multimodal biometric system for smartphones user identification based on multitrait selfie (i.e., of face and ocular regions), user's hand micro-movements, and touchstroke patterns. The proposed system captures *silently* the micro-movement of phone and timing of touch-typing[3] while she is taking face and ocular multitrait selfie and entering a text-independent PIN/password simultaneously in a split-screen mode of smartphones to

[3]Touch-typing is the act of typing input on the touchscreen of a smartphone.

verify the user's identity. To the best of our knowledge, this is the first work that explores combination of face, ocular region, phone-movement, and touchstroke for smartphone user recognition. Use of selfie images enables the proposed system becoming more user-friendly, as acquiring a person's face and ocular region unobtrusively using phone's frontal camera is reasonably easier than acquiring other biometric traits, e.g., fingerprint and palmprint. Moreover, the presented system does not require user to remember any password, graphical pattern, or gesture, since in each attempt user can enter novel text-independent PIN of fixed length. In addition, typing a PIN is easier than writing something on touchscreen (e.g., signature/pattern). Even if impostor knows what is being entered as a PIN, access will still be denied because the system employs keystroke dynamics features (that is specific to each individual) and not the PIN itself, plus impostor cannot mimic the phone micro-movement of genuine user. Also, use of multimodality makes the proposed system more secure than unimodal ones, as spoofing only one or two modalities would not suffice to grant access to the phone [17].

To demonstrate the efficacy of the proposed system, we also collected a multimodal dataset of touchstroke and phone-movement patterns from 95 subjects with multiple smartphones under uncontrolled fashion where user could sit, stand, or walk.[4] Experimental analysis on public MOBIO Face, VISOB ocular, and MultiTouchMove mobile datasets shows promising results with more practical accuracy and usability. Namely, this study shows that multimodal biometrics can be integrated with smartphones in a user-friendly manner with significant improved usability and security.

## 2 Related Work

The face recognition system is a method of identifying a person from an image or a video frame using the user's facial features. This recognition system is extremely popular both in academia and industry because of its accuracy and robustness.

As a unimodal system, face recognition has widely been explored for user authentication on mobile phones [18, 19]. The study [18] introduced a face recognition system for a mobile device having a Symbian Operating System, installed. The approach was based on One-class Support Vector Machines. The authors reported an EER of 7.92% and 3.95% according to the global and individual threshold, respectively. Later, Hadid et al. [19], proposed a face recognition approach using Haar features and Adaboost ensemble classifier. They obtained average authentication accuracy of 82% on small-sized face images (4040 pixels) and 96% for medium size faces (8080 pixels). Similarly, the low-cost face recognition system proposed by Tao and Veldhuis [20] achieved an EER of 2%.

[4]The database is made publicly available for research purposes and can be obtained by contacting the authors.

The face has also been evaluated in bi-modal or tri-modal settings for user authentication on smartphones. The study by Chen et al. [21] introduced a sensor-assisted face recognition scheme that used motion and face image for user authentication. They obtained a 95–97% detection rate at up to 2–3% false alarm rate on 450 test attempts. The study by Zahid et al. [22] proposes a tri-modal system based on face, phone-movements, and touchstrokes. This study profiles users on the unobtrusively collected phone's micro-movements together with the timing differences of entered touchstrokes during the course of users taking selfies. During the evaluation, the authors achieved as much as 99% TAR at 1% EER on the chimerical dataset of their collected touchstrokes plus phone-movements, with the face images from MOBIO Dataset [23]. The scheme has the potential to be used not only for smartphone unlocking but also as separate authentication service for critical applications, like mobile banking. The proposed scheme profiles the user's phone micro-movements and touchstrokes during the course of users capturing their selfies. It later extracts the features from all the modalities and fuses them and prepares the template. Later, it authenticates the users based on the differences in BSIF learned features from the face and ocular, and statistical features learned from the behavioral biometrics.

Earlier work [22] proposed a tri-modal system that used the phone's micro-movements, touchstroke timings, and facial patterns to authenticate the users of the smartphones. This study enhances that scheme by introducing an additional modality—the ocular modality. Additionally, we re-run all the experiments and report the average results obtained on the fused chimerical dataset of all four modalities from 95 users.

The proposed scheme profiles the user's phone micro-movements and touchstrokes during the course of users capturing their selfies. It later extracts the features from all the modalities and fuses them and prepares the template. Later, it authenticates the users based on the differences in BSIF learned features from the face and ocular, and statistical features learned from touchstroke and phone's micro-movements. Obtained experimental results proved our scheme as extremely accurate and robust; since simultaneous spoofing of all the four modalities is extremely difficult, if not impossible.

## 3 Proposed Multitrait Selfie Smartphone User Authentication System

Smartphones are now being used intensively for accessing and storing sensitive personal data, thereby making user authentication an issue of paramount importance. The balance between security and usability however is a challenging task. For instance, according to the survey result reported in [24], the iris and voice biometric traits were ranked highest in terms of perceived security protection, but lowest in terms of usability mainly due to high requirement of user cooperation. Recently, it

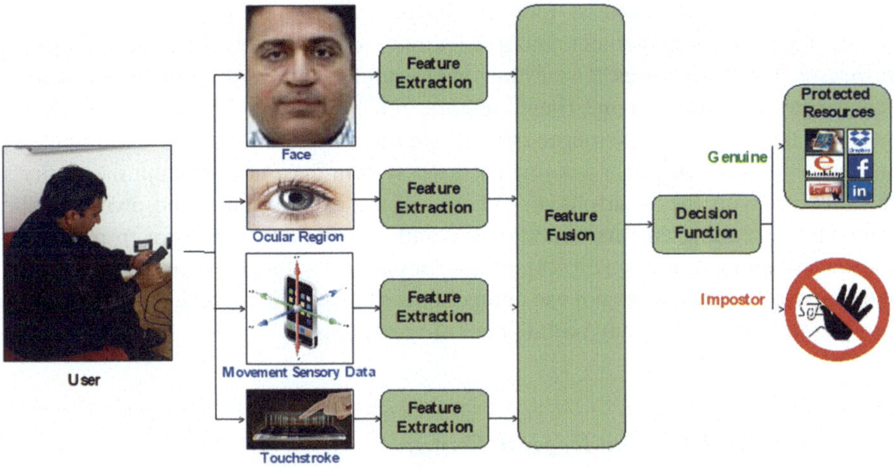


**Fig. 1** Proposed multitrait selfie-based smartphone user authentication framework

has been shown in former studies [5, 8, 9, 25] that each user holds, interacts, and moves her phone in a unique way, which can also be utilized for implicit user authentication. Also, selfie (i.e., a self-portrait photograph) has by now become an epidemic. For instance, a study by Galaxy Research,[5] in 2017, has found that more than 50 million selfies a week are snapped nationally in Australia for various purposes. Indeed, a study by Mitek[6] revealed that 68% of US millennial smartphone users would rather use mobile capture (including selfies) for data extraction and other motives (including authentication), which is easier compared to other form of user authentication. In this work, we propose an unconstrained low-cost multimodal biometric user authentication system based on user's multitrait selfie of face and periocular region, hand micro-movements, and touchstroke patterns.

The proposed framework, as illustrated also in Fig. 1, profiles silently the user's hand micro-movements and touchstrokes simultaneously while user is capturing their multitrait selfie of face and periocular regions and entering 8-digit text-independent passcode in order to access either the phone or any app. The proposed framework, during authentication, works in a split-screen mode of smartphones such that the user concurrently can enter the passcode and see their selfie image being showed and captured. The hand micro-movements pattern analyzed is of a very short period of time, i.e., time taken by the user to enter 8 digits. The rationale behind choosing this time duration is because it was empirically determined that this time is sufficient enough for pattern discrimination and this duration is too short for an adversary to debug the device [26]. The 8-digit passcode was adopted owing to the fact that most apps, document storing/sharing, and social networking services use minimum 8-digit password. Analogous time is also allowed to the user for capturing the multitrait selfie.

[5] http://www.galaxyresearch.com.au/.

[6] https://www.miteksystems.com/.

In particular, the captured four mobile biometric traits (i.e., face, periocular region, hand micro-movement, and touchstroke) are separately processed to extract respective salient features. These four independent features are combined using a feature level fusion scheme (here, feature concatenation). The resultant feature vector is then fed to a classifier to obtain final binary decision: genuine or impostor. If the user is classified as a genuine user, the system does not interrupt the owner's interactions with the smartphone. Otherwise, the system alerts the owner of the phone (e.g., sending an email), and may stealthily isolate the impostor from accessing sensitive functionality [27, 28], or ask for explicit authentication [29, 30]. Since the proposed method does not use components of a specific mobile platform, thus it can be implemented for any mobile operating systems.

## 3.1 Face and Ocular Region Features

In the proposed system, face and ocular region processing sub-systems are composed of two steps: face/periocular region localization and face/periocular region feature extraction. Face and periocular region localization are achieved by the classification and regression tree analysis (CART) based on Haar features [31]. The detected face and periocular region are first normalized, and then Binarized Statistical Image Features (BSIF) technique is applied to obtain respective face and periocular region features. Binarized Statistical Image Features (BSIF) is a local image descriptor constructed by binarizing the responses to linear filters [32]. BSIF learns a set of filters from natural images using an ICA (Independent Component Analysis) based unsupervised scheme. These learned filters are used to represent each pixel of the given image as a binary string by computing its response to learned filters. The binary string for each pixel can be considered as a local descriptor of the image intensity pattern in the neighborhood of that pixel. Finally, the histogram of the pixels binary string values allows one to characterize the texture properties within the image sub-regions.

In this work, we have utilized the open-source filters [32], which were trained using 50,000 image patches randomly sampled from 13 different natural scenic images [33]. Three main steps build the BSIF filters: mean subtraction of each patches, dimensionality reduction using PCA (Principle Component Analysis), and estimation of statistically independent filters (or basis) using ICA. Given a visual laughter sample $I$ of size $l \times m$ and a filter $F_i$ of same size, filter response is attained as follows [32]:

$$r_i = \sum_{l,m} I(l,m) F_i(l,m) \tag{1}$$

Where, $F_i$; $\forall = \{1, 2, \ldots, m\}$ represents statistically independent filters whose response can be together calculated and binarized to obtain the binary string as follows [32]:

$$b_i = \begin{cases} 1, \text{if } r_i > 0 \\ 0, \text{Otherwise} \end{cases} \tag{2}$$

Finally, the BSIF features are obtained as a normalized histogram of pixel's binary codes, which allows one to characterize the texture properties within the face and ocular sub-regions.

## 3.2 Phone-Movements Features

For phone-movements and touchstroke dynamics, the method leverages three built-in three-dimensional sensors: the accelerometer, the gravity sensor, and the magnetometer. Two additional sensor readings are derived from the accelerometer by applying two filters[7] (low pass and high pass) with the parameter $\alpha = 0.5$, named the outcomes Low-Pass Filter (LPF) and High-Pass Filter (HPF) accelerometer readings. Thus, in total, three variants of accelerometer sensor readings: Raw, LPF, and HPF accelerometer readings are obtained. A raw accelerometer reading produces raw values including gravity values. An LPF accelerometer reading measures the apparent transient forces acting on the phone, caused by the user activity, and an HPF reading produces the exact acceleration applied by the user on the phone. The gravity sensor provides the magnitude and direction of the gravity force applied on the phone. The coordinate system and the unit of measurement of gravity sensor are the same as those of the accelerometer sensor. The magnetometer sensor measures the strength and/or direction of the magnetic field in three dimensions. It differs from the compass as it does not provide point north. The magnetometer measures the Earth's magnetic field if the device is placed in an environment absolutely free of magnetic interference. All the above sensors generate continuous streams in $x$, $y$, and $z$ directions. We have added a fourth dimension to all of these sensors and named it *magnitude*. Magnitude has been tested in the context of smartphone user authentication [19, 21, 31] and has proved to be very effective in classification accuracy. The magnitude is mathematically represented as:

$$S_M = \sqrt{\left(a_x^2 + a_y^2 + a_z^2\right)} \tag{3}$$

where $S_M$ is the resultant dimension, and $a_x$, $a_y$, and $a_z$ are the accelerations along the $X$, $Y$, and $Z$ directions.

[7] http://developer.android.com/guide/topics/sensors/sensors.motion.html.

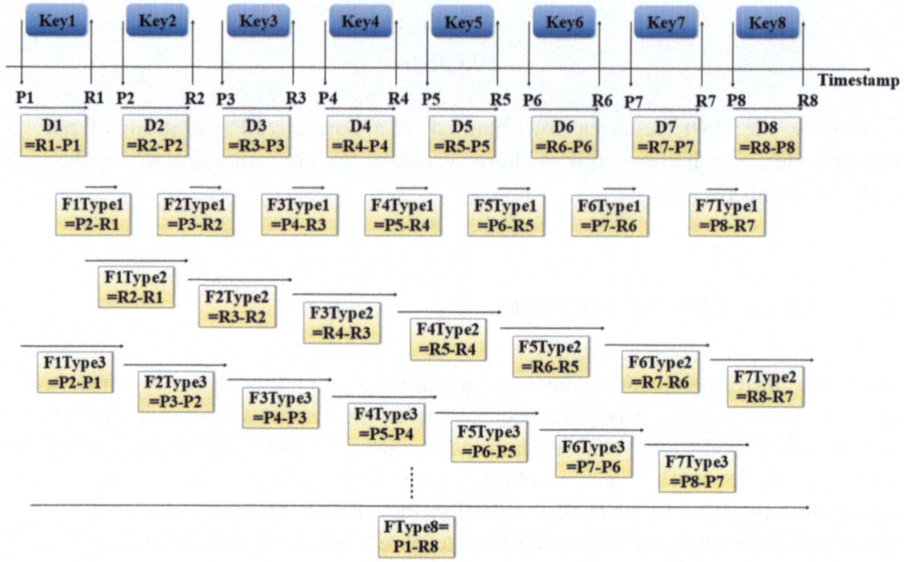


**Fig. 2** Touchstroke features used in this work. $\mathbf{P}_i$ and $\mathbf{R}_i$ ($i$ = 1, …,8) stand for press and release of keys

## 3.3 Touchstroke Features

To extract touchstroke features, the system leverages three built-in three-dimensional sensors and their variants discussed above in Sect. 11.2. In particular, the touchstroke features are extracted using typing *n*-graph, namely dwell time and flight time from each typing pattern, as depicted in Fig. 2.

## 3.4 Classification Methods

Final decision, whether the user is genuine or imposter, is usually performed either with binary classification (training with two classes) or with anomaly detection (training with only the target class). In binary classification, the classifiers learn to discriminate the true user from a given training set, whereas anomaly detectors look for deviation from the legitimate user's behavior to make a decision. One class classifiers (anomaly detectors) are preferred in final proof-of-the-concept implementation, and, the binary class classification approach is used in offline analysis. Since this work is performed offline, Multilayer Perceptron (MLP) and Random Forest (RF) were used, as binary classifiers, due to their simplicity, fast computation, and resistance against overfitting.

# 4 Experiments

Here, we provide an experimental evaluation of the proposed multitrait selfie-based multimodal smartphone user authentication framework.

## 4.1 Datasets

*MOBIO Face database* [23]. This public dataset consists of face samples collected from 150 subjects using a NOKIA N93i mobile phone under realistic and uncontrolled environment. It was captured in phase I and II, which respectively include 21 and 11 videos per user with six sessions. Figure 3 shows examples from the MOBIO database.

*VISOB (Visible Light Mobile Ocular Biometric)* database [34]. It is a large-scale public dataset that was collected in mobile environment under non-controlled sample variations, acquisition, and setups. The VISOB dataset is composed of 550 subjects comprising eye images captured using front facing (selfie) cameras of three different mobile devices, namely Oppo (13 MP), Samsung, and iPhone. For each device, three different subsets are provided that correspond to three different lighting conditions: regular office, dim light (indoors), and bright daylight (outdoors). Each subset contains respective enrollment and (short term) verification sets. This dataset specifically presents possible intra-class variations due to the nature of mobile front facing cameras and everyday mobile biometrics use cases, such as out-of-focus images, occlusions due to prescription glasses, different illumination conditions, gaze deviations, eye-makeup (i.e., eye liner and mascara), specular reflections, and motion blur. Figure 4 shows examples from the VISOB database.

*Movement and Touchstroke database*. We developed a customized Android application for the purpose of this data collection. The app collects data from multiple sensors (i.e., accelerometer, gravity, orientation, magnetometer, and gyroscope sensors) and LPF and HPF values [5]. While Android supports data collection in four fixed delays (often termed as *Sensor_Delay_Modes*): *Sensor_Delay_Normal*, *Sensor_Delay_Game*, *Sensor_Delay_UI*, and *Sensor_Delay_Fastest* with fixed delay of 0.2, 0.02, 0.06, and 0 s, respectively. We observed *Sensor_Delay_Normal* to be too slow to sense the movements, and *Sensor_Delay_UI* and *Sensor_Delay_Fastest* very likely to include the noise during data collection. Thus, we used

**Fig. 3** Examples from MOBIO face database

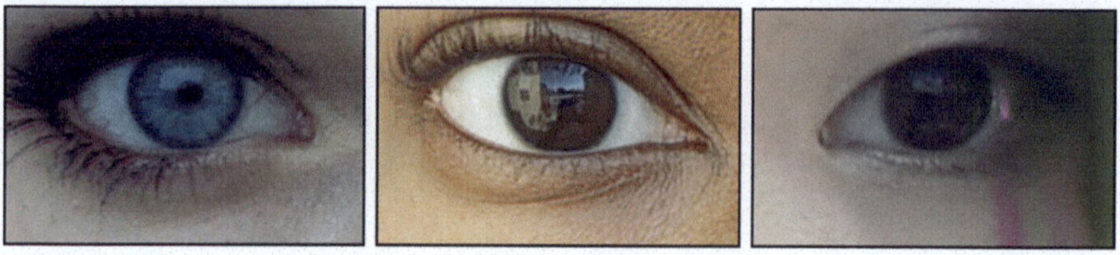

**Fig. 4** Examples of eye images captured using front facing (selfie) cameras of three different mobile devices (L-R): iPhone, Oppo, and Samsung

*Sensor_Delay_Game* in this paper. The data was collected from 95 users under uncontrolled environment using a crowd sourcing platform. The users were required to install the app and enter 8-digit touch-types in different activities, i.e., *sitting*, *standing*, and *walking*. The dataset was collected in three phases (days), 30 samples in each activity per phase from each user. This database is made publicly available by the authors.

## 4.2 Experimental Protocols

Since no multimodal datasets including face, ocular region, phone micro movements, and touckstroke are publicly available, we created a chimerical multimodal datasets by associating the four modalities of pairs of clients of the available *MOBIO*, *VISOB*, and *Movement and Touchstroke* datasets without regard to age and/or gender. Note that building chimerical datasets is a widely used approach in experimental investigations on multimodal biometrics [14]. Following steps were performed on the detected face and ocular regions: conversion to grayscale, histogram equalization, normalization to $175 \times 175$ for face and $150 \times 100$ for ocular trait, and feature extraction using BSIF filter of size $11 \times 11$ with a length of 12 bits. From every 3-dimensional sensor, 4 data streams were collected to extract four statistical features (mean, standard deviation, skewness, and kurtosis) in each data stream. Data from every sensor were then transformed into a $4 \times 4$ features matrix. In total 16 features from all four dimensions of each sensor were obtained. Likewise, 14 features were extracted based on touch-typing timing from the text-independent 8-digit passcode entered by the user. The two adopted classifiers (i.e., MLP and RF) were used from the WEKA workbench [35] for user authentication. The performance was evaluated using TAR (True Acceptance Rate) and EER (Equal Error Rate), which respectively are the fraction of legitimate user attempts classified correctly and where false rejection and false acceptance become equal. To mitigate the class imbalance problem, we used randomly selected $n(= 5 \text{ or } 10)$ and all samples of a genuine user and an impostor, respectively, as training set, whereas remaining samples of the genuine user and all samples of an impostor (that was not used in training) were used as testing set; the testing phase was repeated 94 times using different impostor in each run. Furthermore, all the above procedure was repeated 94 times for the given user using a different impostor in training in each run. Reported results are average values over the 94 × 94 × 95 runs.

## 4.3 Experimental Results and Discussion

We report our obtained results for the proposed multimodal smartphone authentication scheme in 1, 2, 3, and 4 Tables, on 5 and 10 training samples, respectively, under three postures: *sitting*, *standing*, and *walking*.

We could extract several observations from the tables: (1) the proposed authentication scheme is simple yet accurate, on one hand, and applicable in in-the-wild conditions, on another. Additionally, its quality of being implicit and novel makes it more usable and acceptable, among the users. Since, users are not expected to remember any secret (e.g., a PIN or a password), the proposed scheme is highly like to be widely acceptable; (2) RF classifier and phone micro-movement modality, by and large, outperform other considered classifier and modalities, respectively, owing to RF's ability of reducing the variances, averaging out the biases and most unlikeliness of overfitting, and micro-movement modality's very low intra- and high inter-class variation; (3) increasing the number of training samples increases identification accuracy, it however reduces the usability as in practice most users likely train their systems with fewer samples, e.g., during our data collection 49.5% users said they would prefer to use less than or maximum five samples for training the system; (4)

**Table 1** Performance of both classifiers (averaged over all 95 users) in *sitting*, *standing*, and *walking* activities in unimodal settings (on five and ten training samples)

| | | | Movements | | Touch | | Face | | Occular | |
|---|---|---|---|---|---|---|---|---|---|---|
| Activity | Training samples | Classifiers | TAR | EER | TAR | EER | TAR | EER | TAR | EER |
| *Sitting* | *5* | MLP | 0.98 | 0.02 | 0.78 | 0.22 | 0.57 | 0.43 | 0.84 | 0.16 |
| | | RF | 0.97 | 0.03 | 0.82 | 0.18 | 0.52 | 0.48 | 0.76 | 0.24 |
| | *10* | MLP | 0.99 | 0.01 | 0.79 | 0.21 | 0.70 | 0.30 | 0.95 | 0.05 |
| | | RF | 0.99 | 0.01 | 0.86 | 0.14 | 0.54 | 0.46 | 0.88 | 0.12 |
| *Standing* | *5* | MLP | 0.98 | 0.02 | 0.80 | 0.20 | 0.54 | 0.46 | 0.84 | 0.16 |
| | | RF | 0.98 | 0.02 | 0.84 | 0.16 | 0.49 | 0.51 | 0.76 | 0.24 |
| | *10* | MLP | 0.99 | 0.01 | 0.82 | 0.18 | 0.61 | 0.39 | 0.95 | 0.05 |
| | | RF | 0.99 | 0.01 | 0.87 | 0.13 | 0.50 | 0.59 | 0.88 | 0.12 |
| *Walking* | *5* | MLP | 0.98 | 0.02 | 0.81 | 0.19 | 0.58 | 0.42 | 0.84 | 0.16 |
| | | RF | 0.97 | 0.03 | 0.84 | 0.16 | 0.52 | 0.58 | 0.76 | 0.24 |
| | *10* | MLP | 0.99 | 0.01 | 0.84 | 0.16 | 0.66 | 0.34 | 0.95 | 0.05 |
| | | RF | 0.99 | 0.01 | 0.91 | 0.09 | 0.54 | 0.46 | 0.88 | 0.12 |

The results of occular biometrics is same because the chosen dataset was not collected in different activities

**Table 2** Performance of both classifiers (averaged over all 95 users) in *sitting*, *standing*, and *walking* activities in bi-modal settings (on five and ten training samples)

| Activity | Training samples | Classifiers | Movement + Touch | | Movement + Face | | Movement + Occular | | Touch + Face | | Face + Occular | | Touch + Occular | |
|---|---|---|---|---|---|---|---|---|---|---|---|---|---|---|
| | | | TAR | EER | TAR | EER | TAR | EER | TAR | EER | TAR | EER | TAR | EER |
| *Sitting* | *5* | MLP | 0.97 | 0.03 | 0.93 | 0.07 | 0.89 | 0.11 | 0.70 | 0.30 | 0.84 | 0.16 | 0.85 | 0.15 |
| | | RF | 0.97 | 0.03 | 0.94 | 0.06 | 0.88 | 0.12 | 0.60 | 0.40 | 0.76 | 0.24 | 0.79 | 0.21 |
| | *10* | MLP | 0.99 | 0.01 | 0.97 | 0.03 | 0.98 | 0.02 | 0.79 | 0.21 | 0.94 | 0.06 | 0.95 | 0.05 |
| | | RF | 0.99 | 0.01 | 0.97 | 0.03 | 0.97 | 0.03 | 0.70 | 0.30 | 0.86 | 0.14 | 0.91 | 0.09 |
| *Standing* | *5* | MLP | 0.98 | 0.02 | 0.94 | 0.06 | 0.90 | 0.10 | 0.72 | 0.28 | 0.84 | 0.16 | 0.85 | 0.15 |
| | | RF | 0.98 | 0.02 | 0.97 | 0.03 | 0.91 | 0.09 | 0.73 | 0.27 | 0.72 | 0.28 | 0.80 | 0.20 |
| | *10* | MLP | 0.99 | 0.01 | 0.99 | 0.01 | 0.98 | 0.02 | 0.82 | 0.18 | 0.94 | 0.06 | 0.95 | 0.05 |
| | | RF | 0.99 | 0.01 | 0.99 | 0.01 | 0.98 | 0.02 | 0.80 | 0.20 | 0.86 | 0.14 | 0.91 | 0.09 |
| *Walking* | *5* | MLP | 0.98 | 0.02 | 0.94 | 0.06 | 0.89 | 0.11 | 0.76 | 0.24 | 0.83 | 0.17 | 0.83 | 0.17 |
| | | RF | 0.98 | 0.02 | 0.96 | 0.04 | 0.90 | 0.10 | 0.78 | 0.22 | 0.75 | 0.25 | 0.75 | 0.25 |
| | *10* | MLP | 0.99 | 0.01 | 0.98 | 0.02 | 0.98 | 0.02 | 0.82 | 0.18 | 0.94 | 0.06 | 0.95 | 0.05 |
| | | RF | 0.99 | 0.01 | 0.98 | 0.02 | 0.98 | 0.02 | 0.80 | 0.20 | 0.87 | 0.13 | 0.92 | 0.08 |

**Table 3** Performance of both classifiers (averaged over all 95 users) in *sitting*, *standing*, and *walking* activities in Tri-modal settings (on five and ten training samples)

| Activity | Training samples | Classifiers | Movement + Touch + Face | | Movement + Face + Occular | | Movement + Touch + Occular | | Touch + Face + Occular | |
|---|---|---|---|---|---|---|---|---|---|---|
| | | | TAR | EER | TAR | EER | TAR | EER | TAR | EER |
| *Sitting* | *5* | MLP | 0.92 | 0.08 | 0.88 | 0.12 | 0.88 | 0.12 | 0.84 | 0.16 |
| | | RF | 0.95 | 0.05 | 0.87 | 0.13 | 0.89 | 0.11 | 0.79 | 0.21 |
| | *10* | MLP | 0.96 | 0.04 | 0.97 | 0.03 | 0.97 | 0.03 | 0.95 | 0.05 |
| | | RF | 0.97 | 0.03 | 0.97 | 0.03 | 0.98 | 0.02 | 0.89 | 0.11 |
| *Standing* | *5* | MLP | 0.94 | 0.06 | 0.89 | 0.11 | 0.89 | 0.11 | 0.84 | 0.16 |
| | | RF | 0.97 | 0.03 | 0.88 | 0.12 | 0.92 | 0.08 | 0.77 | 0.23 |
| | *10* | MLP | 0.99 | 0.01 | 0.97 | 0.03 | 0.98 | 0.02 | 0.94 | 0.06 |
| | | RF | 0.99 | 0.01 | 0.97 | 0.03 | 0.98 | 0.02 | 0.89 | 0.11 |
| *Walking* | *5* | MLP | 0.94 | 0.06 | 0.88 | 0.12 | 0.89 | 0.11 | 0.83 | 0.17 |
| | | RF | 0.97 | 0.03 | 0.89 | 0.11 | 0.92 | 0.08 | 0.79 | 0.21 |
| | *10* | MLP | 0.98 | 0.02 | 0.97 | 0.03 | 0.98 | 0.02 | 0.94 | 0.06 |
| | | RF | 0.99 | 0.01 | 0.97 | 0.03 | 0.98 | 0.02 | 0.91 | 0.09 |

**Table 4** Performance of both classifiers (averaged over all 95 users) in *sitting*, *standing*, and *walking* activities in 4-Modal settings (on five and ten training samples)

| Activity | Training samples | Classifiers | Movement + Touch + Face + Occular | |
|---|---|---|---|---|
| | | | TAR | EER |
| *Sitting* | *5* | MLP | 0.88 | 0.12 |
| | | RF | 0.89 | 0.11 |
| | *10* | MLP | 0.97 | 0.03 |
| | | RF | 0.97 | 0.03 |
| *Standing* | *5* | MLP | 0.89 | 0.11 |
| | | RF | 0.90 | 0.10 |
| | *10* | MLP | 0.97 | 0.03 |
| | | RF | 0.98 | 0.02 |
| *Walking* | *5* | MLP | 0.88 | 0.12 |
| | | RF | 0.91 | 0.09 |
| | *10* | MLP | 0.97 | 0.03 |
| | | RF | 0.98 | 0.02 |

among adopted modalities face comparatively achieved high error rates[8] mainly because samples in-the-wild generally exhibit low texture clarity (see Fig. 3), plus BSIF hugely depends on ICA that works better for non-Gaussian data, while we

[8] These are our rudimentary results. We are presently developing inexpensive deep learning topologies for face recognition in-the-wild on smartphones for elevated accuracy and interoperability.

observed that face dataset samples in this study tend strongly to be Gaussian; (5) phone micro-movement is the most accurate modality among the four; however, unimodal systems are vulnerable to spoofing. In fact, it has been shown in [17] that spoofing the best individual modality creates serious security breaches, i.e., attacker has higher chances to evade the system when one spoofs most accurate modality and vice versa; (6) integration of two/three modalities is not consistently beneficial. Generally, benefits of fusion are exploited when modalities show complementary nature [7]. However, since in this case, face modality is significantly worse than others, the performance of the multimodal system is thus below the best performing modality. Nonetheless, the performance drop in the tri-modal and 4-model systems is not very much. Moreover, it is worth noticing that the proposed multimodal system (Phone-movement+Touchstroke+Face+Occular) still procures higher accuracy using only 5 or 10 training samples than multimodal methods in [5, 36] using 25 and 30 samples, respectively. All in all, fusion improves security against spoofing since an attacker has lower chances to evade multimodal systems (even if he spoofs all fused traits) than to evade each single unimodal system [17]. Namely, spoofing four modalities simultaneously is much harder than a single modality. Though, face is easiest to spoof, the study [17] has shown that spoofing the least accurate modality leads to a very low probability of evading multimodal biometric systems.

## 5 Conclusion

Smartphones have become pervasive that are being used to access and store sensitive data. Thus, there is a pressing demand for a more secure, low cost, and user-friendly smartphone user authentication solution. This paper presents an unconstrained smartphone multimodal biometric system using face, ocular region, touchstroke, and phone-movement patterns. The proposed architecture identifies the user by taking silently into account micro-movements of phone, movements of user's finger on touchscreen while user is typing and talking multitrait selfie of face and periocular regions features simultaneously in a split-screen mode of smartphones. This study also collected and shares publicly a mobile multimodal dataset of touchstroke and phone-movement patterns in-the-wild from 95 subjects. Experimental results show high accuracy with increased security and usability. Since proposed method is generic it can be implemented on any smartphone to unlock the device and/or a separate authentication service for various applications, e.g., mobile banking. Future work will focus on analyzing vulnerability of the proposed system against spoofing and thereby developing anti-spoofing techniques. Similarly, development of inexpensive deep learning topologies for face and periocular biometric recognition in-the-wild on smartphones will be explored for elevated accuracy and interoperability.

## References


1. Meng, W., Wong, D. S., Furnell, S., & Zhou, J. (2015). Surveying the development of biometric user authentication on Mobile phones. *IEEE Communications Surveys & Tutorials, 17*(3), 1268–1293.
2. Schaub, F., Deyhle, R., & Weber, M. (2012). Password entry usability and shoulder surfing susceptibility on different smartphone platforms. In *Proc. of the 11th Int'l Conf. on Mobile and Ubiquitous Multimedia* (pp. 1–10).
3. Akhtar, Z., Micheloni, C., & Foresti, G. L. (2015). Biometric liveness detection: Challenges and research opportunities. *IEEE Security & Privacy, 13*(5), 63–72.
4. De Luca, A., Hang, A., von Zezschwitz, E., & Hussmann, H. (2015). I feel like I'm taking Selfies all day!: Towards understanding biometric authentication on smartphones. In *Proc. of the 33rd Annual ACM Conference on Human Factors in Computing Systems* (pp. 1411–1414).
5. Buriro, A., Crispo, B., Delfrari, F., & Wrona, K. (2016). Hold and sign: A novel behavioral biometrics for smartphone user authentication. In *IEEE Security and Privacy Workshops* (pp. 276–285). San Jose, CA: IEEE.
6. Harbach, M., von Zezschwitz, E., Fichtner, A., De Luca, A., & Smith, M. (2014). It's a hard lock life: A field study of smartphone (un)locking behavior and risk perception. In *Symposium On Usable Privacy and Security* (pp. 213–230).
7. Maltoni, D., Maio, M., Jain, A. K., & Prabhakar, S. (2009). *Handbook of Fingerprint Recognition* (2nd ed.). London: Springer.
8. Sitov, Z., et al. (2016). HMOG: New Behavioral Biometric Features for Continuous Authentication of Smartphone Users. *IEEE Transactions on Information Forensics and Security, 11*(5), 877–892.
9. Zhu, J., Wu, P., Wang, X., & Zhang, J. (2013). Sensec: Mobile security through passive sensing. In *IEEE Int'l Conf. on Computing, Networking and Communications (ICNC)* (pp. 1128–1133).
10. Buriro, A., Crispo, B., DelFrari, F., & Wrona, K. (2015). Touchstroke: Smartphone user authentication based on touch-typing biometrics. In *New Trends in Image Analysis and Processing–ICIAP Workshops* (pp. 27–34). Cham: Springer.
11. Sae-Bae, N., Ahmed, K., Isbister, K., & Memon, N. (2012). Biometric-rich gestures: A novel approach to authentication on multi-touch devices. In *Proc. of Conf. on Human Factors in Computing Systems* (pp. 977–986).
12. Blanco-Gonzalo, R., Sanchez-Reillo, R., Miguel-Hurtado, O., & Liu-Jimenez, J. (2013). Performance evaluation of handwritten signature recognition in mobile environments. *IET Biometrics, 3*(3), 139–146.
13. Bhanu, B., & Govindaraju, V. (2011). *Multibiometrics for human identification*. Cambridge: Cambridge University Press.
14. Jain, A. K., Flynn, P., & Ross, A. (2008). *Handbook of multibiometrics*. Cham: Springer.
15. Kisku, D. R., Rattani, A., Gupta, P., & Sing, J. K. (2009). Biometric sensor image fusion for identity verification: A case study with wavelet-based fusion rules graph matching. In *IEEE Conference on Technologies for Homeland Security* (pp. 433–439).
16. Akhtar, Z., Buriro, A., Crispo, B., & Falk, T. (2017). Multimodal smartphone user authentication using Touchstroke, phone-movement and face patterns. In *IEEE Global Conference on Signal and Information Processing (GlobalSIP)* (pp. 1368–1372).
17. Akhtar, Z., (2012). *Security of multimodal biometric systems against spoof attacks*. PhD thesis, University of Cagliari, Italy.
18. Abeni, P., Baltatu, M., & D'Alessandro, R. (2006). *Implementing biometrics based authentication for mobile devices* (pp. 1–5). San Francisco, CA: IEEE Globecom.
19. Hadid, A., Heikkila, J. Y., Silvén, O., & Pietikainen, M. Face and eye detection for person authentication in mobile phones. In *First ACM/IEEE International Conference on Distributed Smart Cameras* (pp. 101–108).

20. Tao, Q., & Veldhuis, R. (2010). Biometric authentication system on mobile personal devices. *IEEE Transactions on Instrumentation and Measurement, 59*, 763–773.
21. Chen, S., Pande, A., & Mohapatra, P. (2014). Sensor-assisted facial recognition: An enhanced biometric authentication system for smartphones. In *Proceedings of the 12th Annual International Conference on Mobile Systems, Applications, and Services* (pp. 109–122).
22. Akhtar, Z., Buriro, A., Crispo, B., & Falk, T. H. (2017). Multimodal smartphone user authentication using touchstroke, phone-movement and face patterns. In *Proceedings of the IEEE Global Conference on Signal and Information Processing (GlobalSIP)* (pp. 1368–1372).
23. McCool, C., Marcel, S., Hadid, A., Pietikinen, M., Matejka, P., Cernock, J., Poh, N., Kittler, J., Larcher, A., Lvy, C., Matrouf, D., Bonastre, J., Tresadern, P., & Cootes, T. (2012). Bi-Modal Person Recognition on a Mobile Phone: using mobile phone data. In *IEEE ICME Workshop on Hot Topics in Mobile Multimedia* (pp. 1–6).
24. Teh, P. S., Zhang, N., Jin Teoh, A. B., & Chen, K. (2016). A survey on touch dynamics authentication in mobile devices. *Computers & Security, 59*(2), 210–235.
25. Buriro, A., Crispo, B., Del Frari, F., Klardie, J., & Wrona, K. (2014). ITSME: Multi-modal and unobtrusive behavioural user authentication for smartphones. In *International conference on passwords* (pp. 45–61).
26. Vidas, T., Votipka, D., & Christin, N. (2011). All your droid are belong to us: A survey of current android attacks. In *USENIX Conf. on Offensive Technologies* (pp. 1–10).
27. Shi, W., Yang, J., Jiang, Y., Yang, F., & Xiong, Y. (2011). SenGuard: Passive user identification on smartphones using multiple sensors. In *Proc. of the 2011 IEEE 7th International Conference on Wireless and Mobile Computing, Networking and Communications* (pp. 141–148).
28. Zhauniarovich, Y., Russello, G., Conti, M., Crispo, B., & Fernandes, E. (2014). MOSES: Supporting and enforcing security profiles on smartphones. *IEEE Transactions on Dependable and Secure Computing, 11*(3), 211–223.
29. Riva, O., Qin, C., Strauss, K., & Lymberopoulos, D. (2012). Progressive authentication: Deciding when to authenticate on mobile phones. In *Proc. of the 21st USENIX Conference on Security Symposium* (pp. 1–15).
30. Hayashi, E., Das, S., Shahriyar, S., Hong, J., & Oakley, I. (2013). CASA: Context-aware scalable authentication. In *Symposium on Usable Privacy and Security* (pp. 1–10).
31. Lienhart, R., Kuranov, A., & Pisarevsky, V. (2003). Empirical analysis of detection cascades of boosted classifiers for rapid object detection. In *Proc. of the 25th DAGM Symposium on Pattern Recognition* (pp. 297–304).
32. Kannala, J., & Rahtu, E. (2012). BSIF: Binarized statistical image features. In *International Conference on Pattern Recognition (ICPR)* (pp. 1363–1366).
33. Hyvearinen, A., Hurri, J., & Hoyer, P. O. (2009). *Natural image statistics* (Vol. 39). Berlin: Springer-Verlag.
34. Rattani, A., Derakhshani, R., Saripalle, S. K., & Gottemukkula, V. (2016). ICIP 2016 competition on mobile ocular biometric recognition. In *IEEE International Conference on Image Processing (ICIP)* (pp. 320–324).
35. Holmes, G., et al. (1994). Weka: A machine learning workbench. In *Proc. Australian and New Zealand Conference on Intelligent Information Systems* (pp. 357–361).
36. Shahzad, M., Liu, A. X., & Samuel, A. (2016). Behavior based human authentication on touch screen devices using gestures and signatures. *IEEE Transactions on Mobile Computing, 15*(12), 1–14.
37. Zhauniarovich, Y., Russello, G., Conti, M., Crispo, B., & Fernandes, E. (2014). MOSES: Supporting and enforcing security profiles on smartphones. *IEEE Transactions on Dependable and Secure Computing, 11*(3), 211–223.
38. Riva, O., Qin, C., Strauss, K., & Lymberopoulos, D. (2012). Progressive authentication: Deciding when to authenticate on Mobile phones. In *Proc. of the 21st USENIX Conference on Security Symposium* (pp. 1–15).

39. Sitova, Z., Sedenka, J., Yang, Q., Peng, G., Zhou, G., Gasti, P., et al. (2015). *HMOG: A new biometric modality for continuous authentication of smartphone users*. arXiv preprint arXiv:1501.01199.
40. Buschek, D., De Luca, A., & Alt, F. (2015). Improving accuracy, applicability and usability of keystroke biometrics on mobile touchscreen devices. In *33rd Annual ACM Conference on Human Factors in Computing Systems* (pp. 1393–1402).

# Research of Biometric Characteristics of the Shape of the Ears Based on Multi-Coordinate Methods

**Ruslan Motornyuk, Andrii Bilan, and Stepan Bilan**

## 1 Introduction

The biometric identification of a person by the geometric shape of the ears is currently receiving much attention. The geometric shape of the auricle consists of a number of anatomical components and has a relatively simple structure. Often, the geometric shape of the ear are considered as part of the face when a biometric person identification by face image is carried out. However, the shape of the auricle in each person has its own characteristics that distinguish each person from other people. Therefore, biometric characteristics of the ear are potentially important.

Ears have several advantages over other parts of the body such as:

- more even color distribution;
- immobility of the individual parts of the ear;
- the ear image is small;
- the ability to get an image of the human ear at a considerable distance from the camcorder.

There are many works that confirm the effectiveness of ear biometrics to automate identification and authentication [1–5]. However, there is no clear evidence that the ears are unique. Therefore, it is best to use ear biometry as an additional

R. Motornyuk
PU "Kiev Department" Branch of the Main Information and Computing Center of the JSC "Ukrzaliznytsya", Kyiv, Ukraine
e-mail: xehap0@sw.uz.gov.ua

A. Bilan
The Municipal Educational Institution Mayakskaya Secondary School, Mayak, Moldova

S. Bilan (✉)
State University of Infrastructure and Technology, Kyiv, Ukraine
e-mail: bilan_sm@gsuite.duit.edu.ua



S. Bilan et al. (eds.), *Biometric Identification Technologies Based on Modern Data Mining Methods*, https://doi.org/10.1007/978-3-030-48378-4_12

biometric parameter in multimodal biometric systems. For example, a biometric analysis of the geometric shape of the ear is carried out when it is possible to obtain images of a person's face only in profile and it is not possible to use additional scanning devices to analyze other biometric characteristics.

At the same time, the implementation of the process of automating the biometric identification of a person by the geometric shape of the auricle rests on a number of problems that require solving [6–10]. One of the main problems is getting a clear, separate image of only the auricle. To obtain a clear image of only the shape of the ear, additional means must be used to cut off the image of other parts of the body (hair, neck, etc.). However, this creates significant inconvenience.

If the image is received from a video camera, which is located at a certain distance, then the image includes not only the ear, but also other parts of the body adjacent to it. In this case, the task is to automatically select the auricle in the image. This task is quite complex and can lead to false ear selecting in the image. Therefore, there is the task of improving the accuracy of biometric identification with fuzzy allocation of the auricle in the image.

This chapter describes methods for biometric identification of a person by the geometric shape of the ear with a clear selection of the auricle during the initial scan and with a fuzzy allocation of the auricle. A method for selecting the auricle based on determining the average value, which indicates the location of the scanning window and its size, is described.

## 2 Software Implementation of Biometric Identification of the Geometry of the Auricle with a Clear Selection

A clear selection of the image of the auricle greatly simplifies the method of biometric identification and the means that implement it. A clear image contains only the image of the ear against the selected color background (for example, a white background). Such an image can be obtained using special devices that can be based on different physical principles. It is desirable that they do not give color reflections and shadows. Examples of clearly selected images of the auricle on Fig. 1 are presented.

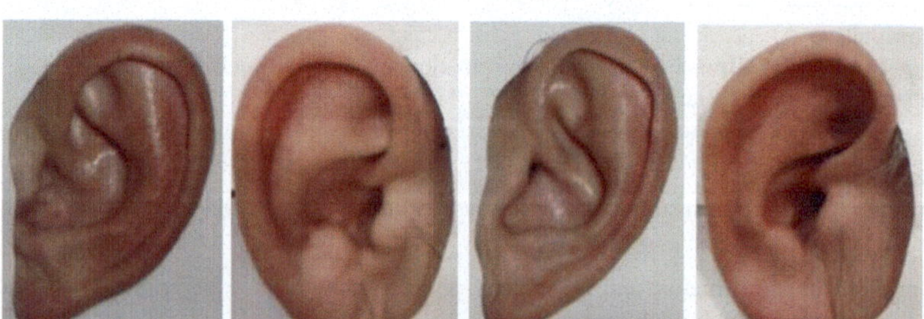

**Fig. 1** Examples of clearly selected images of the auricle

To date, many methods have been developed for analyzing biometric characteristics from the image of the human ear. Basically, all methods use the search for control points and analyze their location and distance between them. The developer or user of the system carries out the choice of control points. These methods use complex algorithms and, accordingly, complex software implementation.

In this paper, we propose a method that is based on the analysis of the shape of the ear using the analysis of individual fragments of the image of the ear. The method is as follows.

1. A clear image of the auricle is formed on a white background using special tools.
2. The resulting image with a selected threshold is binarized.
3. Surrounding rectangle for a binary image of the auricle is formed.
4. Horizontal main line passing through the widest part of the image is drawn and the formed rectangle into upper and lower parts relative to the main line is divided.
5. The bottom of the rectangle is halved with help of an additional horizontal line.
6. The main rectangle into the right and left parts using a vertical line that passes through the vertical line of the image with the most black pixels, is divided.
7. The area of the entire image of the ear and the area of the selected six images are determined.
8. The relationship between the total area of the ear image and the areas of each of the selected six ear images is calculated.
9. The lengths of the main and additional horizontal lines, and also the relationship between them are calculated.
10. Projections of the Radon transform (RT) for each ear are calculated and the relationships between them are calculated [11–13].
11. The obtained quantitative values are compared with the reference values in the database.
12. The system identifies or not the person.

The ninth step of the algorithm is optional and increases the accuracy and reliability of identification. A graphical representation of the identification process by the geometric shape of the ear on Fig. 2 is shown.

The border of the ear is a rectangle describing the ear. Ear boundaries are calculated using four double loops. These loops search for the closest ear pixels to the top, bottom, left, and right edges of the image. As soon as the border is found, the double loop is interrupted.

A fragment of the listing:

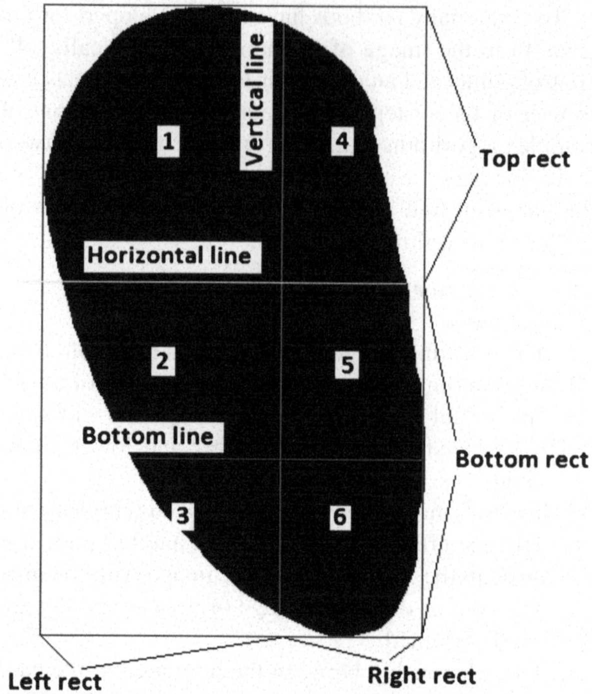


**Fig. 2** Graphic display of the process of personal identification based on the analysis of the area of the image of the ear

```
Main_Rect.Top := -1;
For i := 0 to m - 1 do
 begin
  For j := 0 to n - 1 do
   if M1[j, i] = 1 then
    begin
     Main_Rect.Top := i - 1;
     Break;
    end;
  if Main_Rect.Top > 0 then
   Break;
 end;
Main_Rect.Bottom := -1;
For i := m - 1 downto 0 do
 begin
  For j := 0 to n - 1 do
   if M1[j, i] = 1 then
    begin
     Main_Rect.Bottom := i + 1;
     Break;
    end;
  if Main_Rect.Bottom > 0 then
   Break;
 end;
Main_Rect.Left := -1;
For i := 0 to n - 1 do
 begin
  For j := 0 to m - 1 do
   if M1[i, j] = 1 then
    begin
     Main_Rect.Left := i - 1;
     Break;
    end;
  if Main_Rect.Left > 0 then
   Break;
 end;
Main_Rect.Right := -1;
For i := n - 1 downto 0 do
 begin
  For j := 0 to m - 1 do
   if M1[i, j] = 1 then
    begin
     Main_Rect.Right := i - 1;
     Break;
    end;
  if Main_Rect.Right > 0 then
   Break;
 end;
```

To search and measure the vertical line, a horizontal cycle is organized, inside of which two vertical cycles. Loops vertically bypass the pixels of the bitmap above and below. As soon as the cycles vertically reach the ear above and below they are interrupted. The height of the ear at that location is calculated by the difference Y coordinates. The value of the found ear height is compared with the previous vertical size. If the found value is larger, then the vertical size is assigned the value of the found ear height and the coordinates of the vertical line are assigned new found values.

A fragment of the listing:

```
Line_Vert_1.X := 0;
Line_Vert_1.X := 0;
Line_Vert_2.Y := 0;
Line_Vert_2.Y := 0;
Line_Vert_Size := 0;
For i := 0 to n - 1 do
 begin
  j1 := 0;
  while (M1[i, j1] <> 1) and (j1 < m - 1) do
     Inc(j1);
  j2 := m - 1;
  while (M1[i, j2] <> 1) and (j2 > 0) do
    Dec(j2);
  if (j2 - j1 > Line_Vert_Size) and (j2 - j1 > 0) then
    begin
      Line_Vert_Size := j2 - j1;
      Line_Vert_1.X := i;
      Line_Vert_1.Y := j1;
      Line_Vert_2.X := i;
      Line_Vert_2.Y := j2;
    end;
 end;
```

To search and measure the main horizontal line, a vertical cycle is organized, inside of which two horizontal cycles. The cycles horizontally bypass the pixels of the bitmap on the left and right. As soon as the cycles horizontally reach the ear on the left and right, they are interrupted. Using the difference in X coordinates, the ear width at that location is calculated. The value of the found ear width is compared with the previous horizontal size. If the found value is greater, then the horizontal size is assigned the value of the found ear width and the coordinates of the horizontal line are assigned new found values.

A fragment of the listing:

```
Line_Hor_1.X := 0;
Line_Hor_2.X := 0;
Line_Hor_1.Y := 0;
Line_Hor_2.Y := 0;
Line_Hor_Size := 0;
For i := 0 to m - 1 do
 begin
  j1 := 0;
  while (M1[j1, i] <> 1) and (j1 < n - 1) do
    Inc(j1);
  j2 := n - 1;
  while (M1[j2, i] <> 1) and (j2 > 0) do
   Dec(j2);
  if (j2 - j1 > Line_Hor_Size) and (j2 - j1 > 0) then
   begin
    Line_Hor_Size := j2 - j1;
    Line_Hor_1.X := j1;
    Line_Hor_1.Y := i;
    Line_Hor_2.X := j2;
    Line_Hor_2.Y := i;
   end;
 end;
```

The bottom horizontal line is the line that runs parallel to the horizontal axis X through the Y coordinate, which is in the middle between the Y coordinate of the main horizontal line and the Y coordinate of the lower border of the ear. To do this, first, the Y coordinate is calculated. Then, at the level of this Y coordinate, two horizontal cycles are performed that bypass the pixels of the bitmap on the left and right. As soon as the cycles horizontally reach the ear on the left and right, they are interrupted. The bottom line is assigned the found X coordinate values.

A fragment of the listing:

```
 Line_Bot_1.Y := Line_Hor_1.Y + (Main_Rect.Bottom -
Line_Hor_1.Y) div 2;
 Line_Bot_2.Y := Line_Bot_1.Y;
 j1 := 0;
 while M1[j1, Line_Bot_1.Y] <> 1 do
   Inc(j1);
 Line_Bot_1.X := j1;
 j2 := n - 1;
 while M1[j2, Line_Bot_1.Y] <> 1 do
  Dec(j2);
 Line_Bot_2.X := j2;
```

The dataset of standards used the maximum possible number of reference values for one person. A confidence interval was chosen within which a high accuracy of identification was achieved.

To carry out preliminary preparation of the image of the auricle, the Radon transform was used, which was also used to analyze images of the auricle. The relations between the obtained six projections of the Radon transform, which form additional quantitative values of the characteristic features, were estimated. Figure 3 shows an example of the formation of projections of the RT for all images extracted from images of the auricle.

## 3 Experiment

An experiment was conducted to study the proposed method. Initially, a database was created that contained images of the ears, divided into ten images for each identifiable. Each image of the ear was formed using a video camera located in relation to the auricle at a certain angle. This allowed us to form images of the ears with the same location. Means were also used to isolate only the auricle and cut off the remaining parts of the body. Examples of images of the ears in Fig. 1 are presented.

For the experiment, ten characteristic features are used to the analysis, which were represented by the relations between the selected regions of the auricle and between additional geometric characteristics.

1. The ratio of the total area of the image of the auricle to the first area bounded by the upper left rectangle (Fig. 2).
2. The ratio of the total area of the image of the auricle to the second area bounded by the middle left rectangle.
3. The ratio of the total area of the image of the auricle to the third area bounded by the lower left rectangle.
4. The ratio of the total area of the image of the auricle to the fourth area bounded by the upper right rectangle.
5. The ratio of the total area of the image of the auricle to the fifth area bounded by the middle right rectangle.
6. The ratio of the total area of the image of the auricle to the sixth area bounded by the lower right rectangle.
7. The ratio of the upper area to the lower, separated by a main horizontal line passing through the widest part of the auricle.
8. The ratio of the left area to the right, separated by a vertical line passing through the highest part of the auricle.
9. The ratio of the length of the vertical line to the length of the horizontal line.
10. The ratio of the length of the main horizontal line to the length of the second (lower) horizontal line.

Each image of the auricle was fed to the input of a biometric identification system, the corresponding geometric characteristics were extracted and formed, and their quantitative values were calculated. Based on the obtained quantitative values, the relationships between them described above were calculated.

The choice of a standard was carried out according to two parameters.

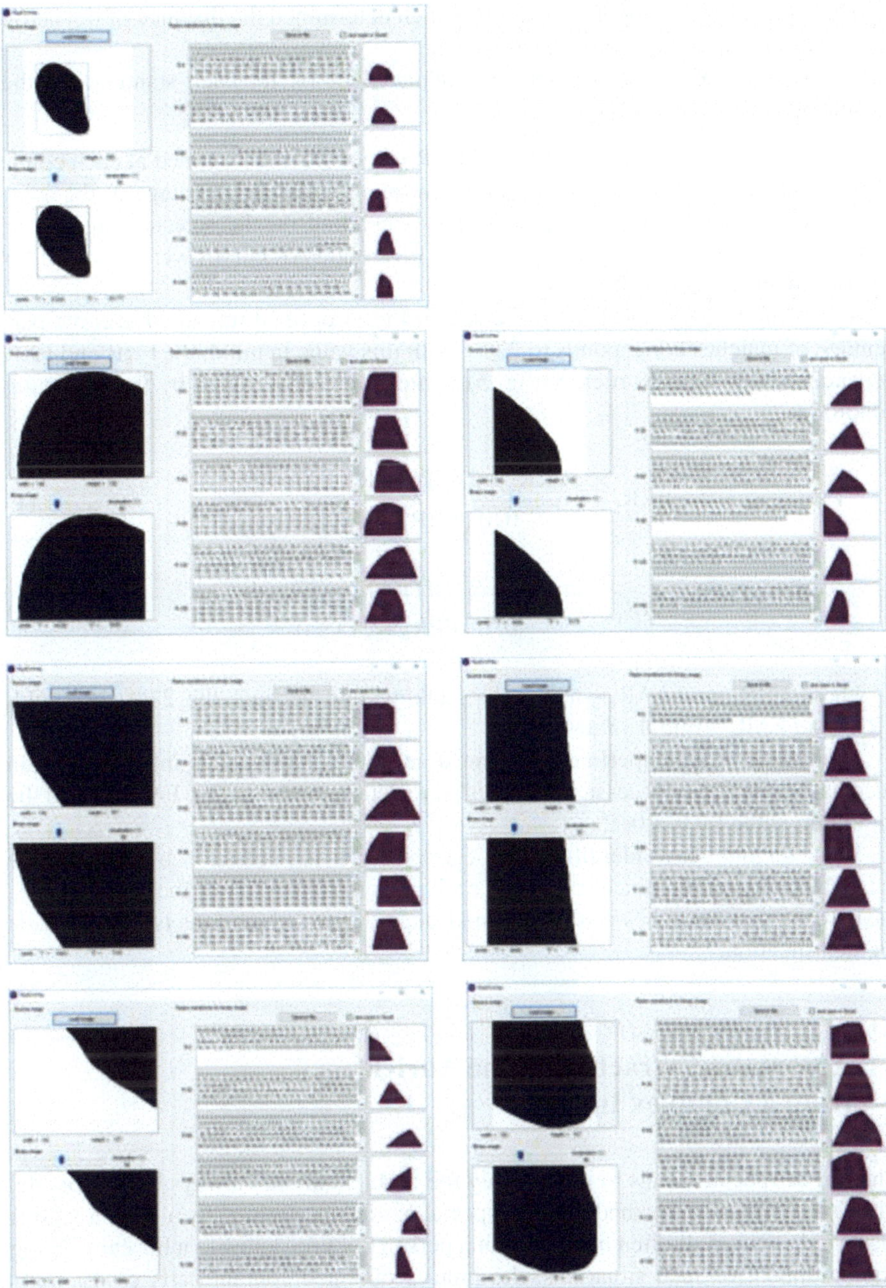

**Fig. 3** RT projection example for selected auricle images

1. The value of the confidence interval, which determined the distance of deviation of the input quantity from the reference.
2. The number of matches from ten calculated ratios for each standard on the selected confidence interval.

During the experiment, the FRR and FAR values were determined at each confidence interval. The FRR and FAR curves were plotted as a percentage of the average value of the identification performed, depending on the selected number of matches (0 ÷ 10). In this case, graphs of the FRR and FAR curves were obtained for each confidence interval (Fig. 4).

According to FRR and FAR presented in Fig. 4 optimal threshold average percentage of matches corresponds to 27%. With this value in mind, the FRR and FAR dependencies were constructed (Fig. 5), which were determined by the following conditions.

$$FRR_1 = \begin{cases} FRR_0 - 27, \text{if}\left(FRR_0 - 27\right) > 0 \\ 0, \text{if}\left(FRR_0 - 27\right) \leq 0 \end{cases}, \tag{1}$$

$$FAR_1 = \begin{cases} FAR_0 - 27, \text{if}\left(FAR_0 - 27\right) > 0 \\ 0, \text{if}\left(FFR_0 - 27\right) \leq 0 \end{cases}, \tag{2}$$

where $FRR_1$ *and* $FAR_1$—new values taking into account the 27% threshold; $FRR_0$ *and* $FAR_0$—initial values.

As a result of the experiment, the most acceptable confidence intervals and an acceptable number of necessary matches were determined that give FAR = 0 and the minimum value FRR = 0÷2.

Thus, for biometric identification of a person, it is necessary to use from the database up to ten reference values for each person. As a result of the comparisons, the percentage of matches was determined, according to which a decision was made on identification.

## 4 Automatic Extraction of the Auricle from a Complex Image

The use of special means to isolate only the auricle may not always be acceptable. Most often, images obtained without special actions performed by a person who is identified are used. It often happens that a person does not know that his ear is being photographed. In such situations, the images obtained contain parts of the body adjacent to the auricle (Fig. 6). These images were freely available from Esther Gonzalez.

Such images require additional complex algorithms that clearly distinguish the auricle from the rest of the background. Images have a wide range of brightness and

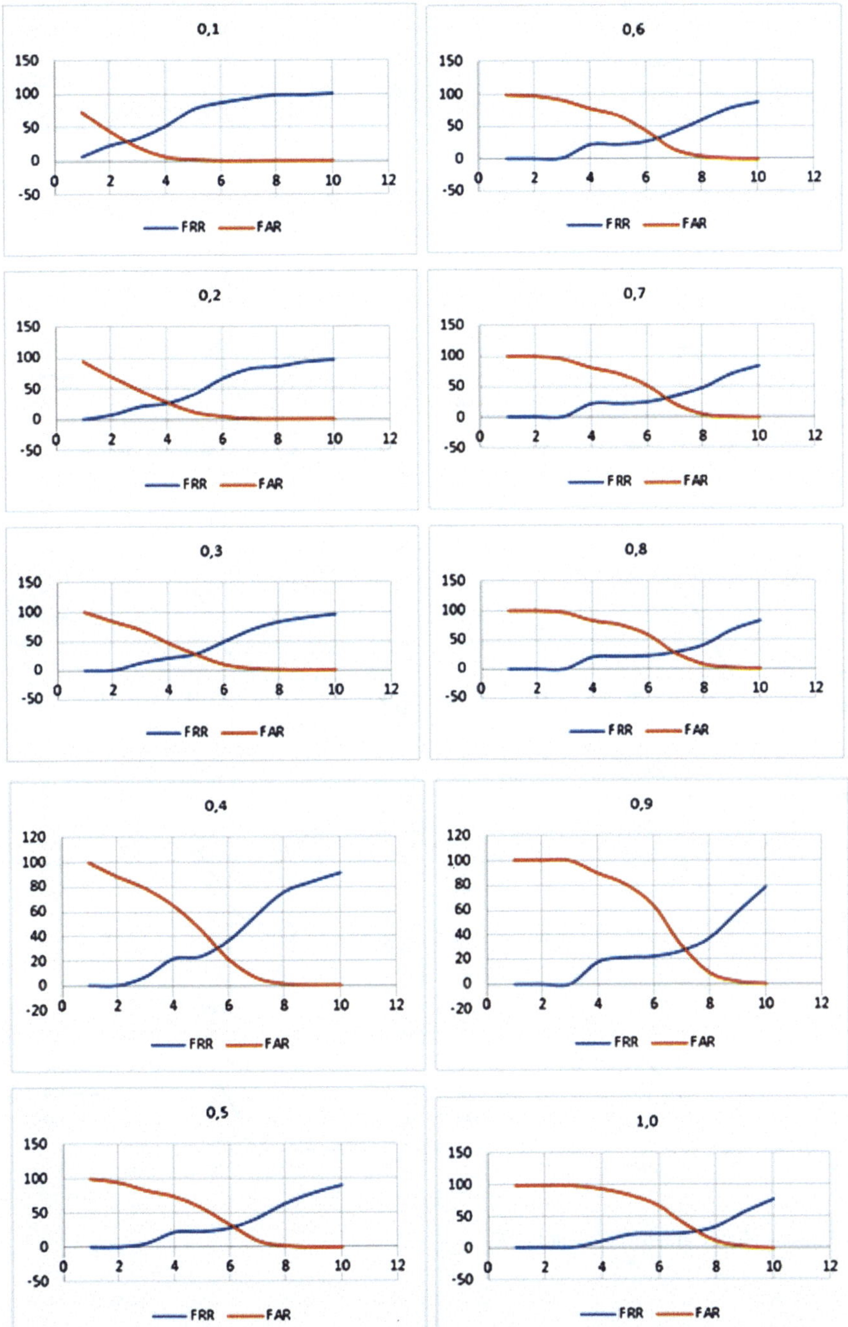


**Fig. 4** FRR and FAR dependencies without a selected percentage identification threshold

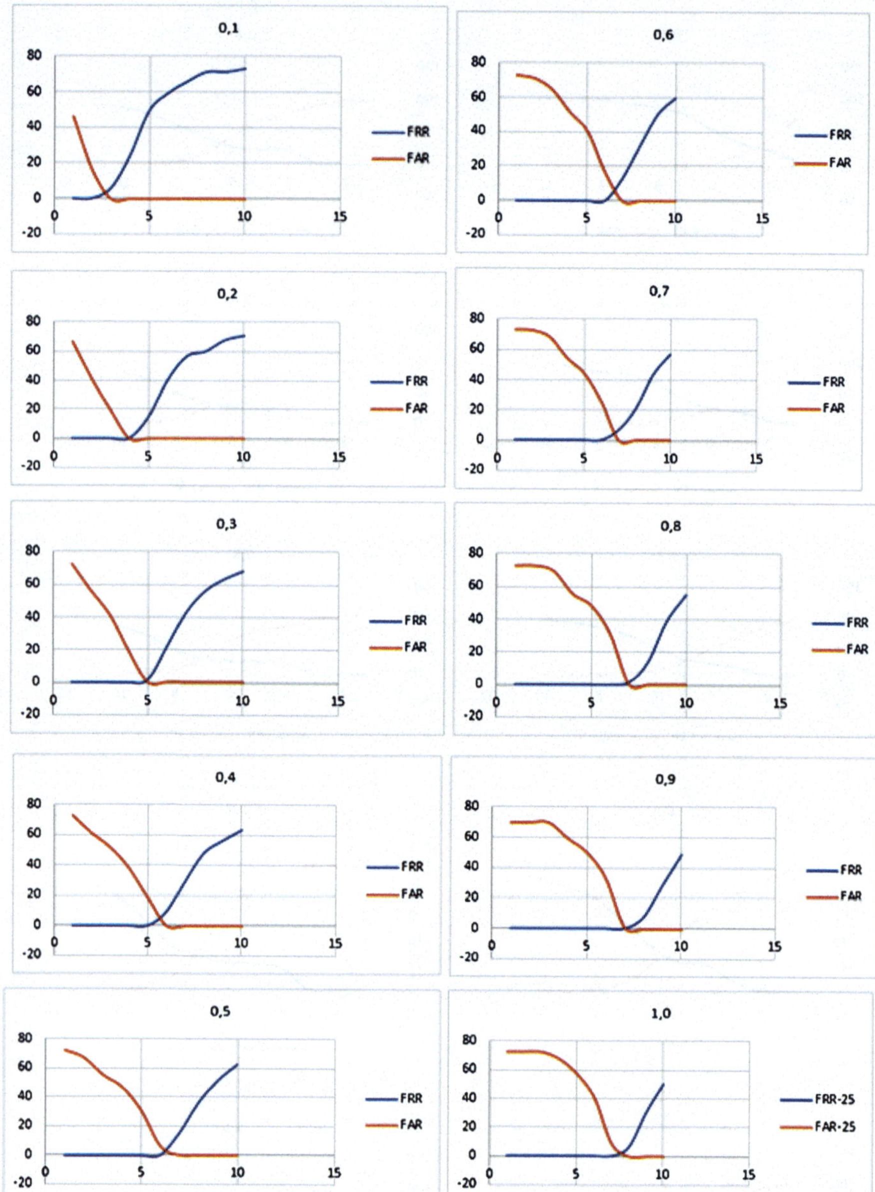


**Fig. 5** Dependency graphs $FRR_1$ *and* $FAR_1$

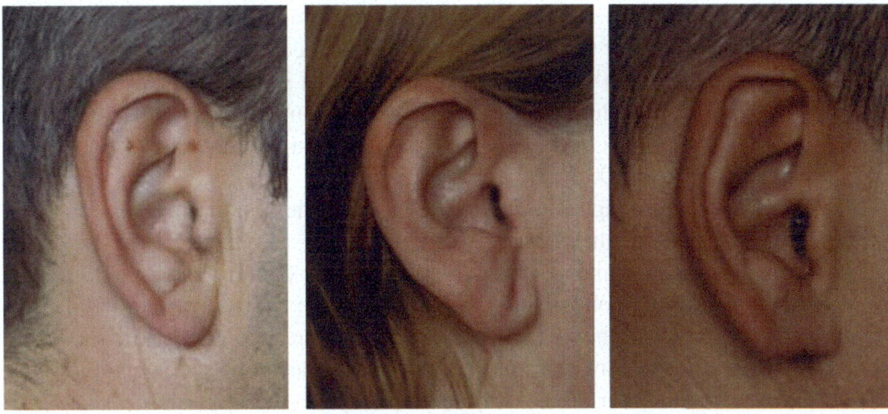

**Fig. 6** Images of ears that were used for the experiment

| | | | | | | | | | | |
|---|---|---|---|---|---|---|---|---|---|---|
| 4145480 | 4276551 | 4277321 | 5001300 | 5725024 | 5198426 | 3816774 | 3356222 | 3421758 | 3882566 | 447470 |
| 4013895 | 4736847 | 5132373 | 5330009 | 5132632 | 4080203 | 3224895 | 3290944 | 3356224 | 3553346 | 388256 |
| 4605519 | 5065813 | 5132375 | 4869461 | 4146508 | 3422788 | 3291202 | 3291202 | 3027774 | 3224896 | 381703 |
| 4737362 | 4605262 | 4408654 | 4474447 | 3883081 | 3160130 | 3094081 | 3028288 | 2962239 | 3554119 | 454101 |
| 4342604 | 4276297 | 4145482 | 4145483 | 3685445 | 2896702 | 2699582 | 2831168 | 3225412 | 4146257 | 519894 |
| 3947845 | 4276297 | 4211275 | 3685444 | 3028027 | 2567992 | 2699836 | 3423559 | 4343891 | 5066847 | 565923 |
| 4079688 | 4342860 | 3751237 | 2962234 | 2633783 | 2897213 | 3620935 | 4738905 | 5659237 | 5724775 | 572477 |
| 4869204 | 4408653 | 2830648 | 2567733 | 3160125 | 3884105 | 5331295 | 5725797 | 5527907 | 5461347 | 539580 |
| 5000791 | 3685187 | 2633010 | 3225661 | 3884104 | 5133657 | 6120296 | 5659748 | 4804183 | 4934744 | 480341 |
| 4277323 | 3356220 | 3224890 | 3883335 | 4870487 | 5725283 | 5528160 | 4935768 | 4475218 | 4474962 | 440891 |
| 3619647 | 3619391 | 4014407 | 4606546 | 5395806 | 5330270 | 4935256 | 4672084 | 4409168 | 4343631 | 447470 |
| 3422269 | 4014407 | 4803412 | 5066584 | 5263964 | 5263964 | 4803413 | 4737620 | 4606034 | 4474448 | 500053 |
| 3685185 | 4145738 | 4737619 | 5000792 | 5132636 | 5263964 | 4671828 | 4540242 | 4474448 | 4671827 | 552713 |
| 3750980 | 4145482 | 4540498 | 4869722 | 5198430 | 4869464 | 4474449 | 4277070 | 4211276 | 4934999 | 598768 |
| 3882567 | 4606034 | 4869464 | 4672343 | 4606294 | 4540756 | 4474449 | 4277327 | 4342863 | 5592929 | 657982 |
| 4474705 | 4935514 | 4606549 | 3948877 | 3949135 | 4409684 | 4211791 | 3817032 | 4803414 | 6250859 | 657982 |
| 5001563 | 4672599 | 3883083 | 3619912 | 3883341 | 3949391 | 3817291 | 4080463 | 5527394 | 6185066 | 579056 |
| 5067357 | 4212305 | 3620168 | 3751754 | 3817547 | 3291720 | 3423305 | 4804701 | 5264736 | 5000792 | 526422 |
| 4343891 | 4146512 | 4014926 | 3752010 | 3028801 | 3029061 | 4344664 | 5199973 | 5001564 | 5000792 | 559292 |
| 3554889 | 4015439 | 4146767 | 3686216 | 3160386 | 3884882 | 5134950 | 5265765 | 5462628 | 5987943 | 585610 |
| 3160130 | 3752009 | 3949131 | 4146251 | 4541267 | 5331298 | 5528677 | 5396579 | 5856873 | 6185579 | 592189 |
| 3291716 | 3291457 | 3685699 | 4343370 | 5592667 | 6710890 | 6382185 | 5527648 | 5791078 | 5659492 | 585661 |
| 3226180 | 3225921 | 3619393 | 4211784 | 5855838 | 7105391 | 6579049 | 6053735 | 5330270 | 5198941 | 631716 |
| 2963008 | 3291458 | 4014408 | 4869458 | 6645612 | 6842476 | 6118757 | 6053736 | 5659236 | 5725285 | 631741 |

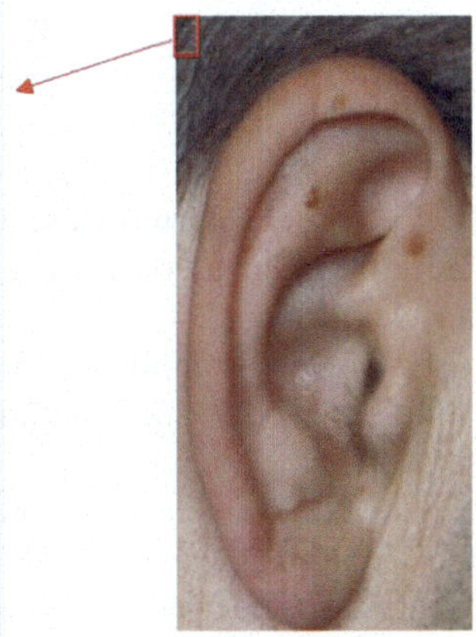

**Fig. 7** An example of a fragment of a two-dimensional array of numbers describing the presented image

color parameters, so sometimes the automatic extraction of the auricle leads to a false result.

An algorithm is proposed in the work, which consists in the following.

The image from the video camera is presented in the form of a two-dimensional array of numbers, each number of which encodes the color and brightness of the corresponding image pixel (Fig. 7). On Fig. 7 is shown only the initial array of size $10 \div 10$ of the whole image.

The system "knows" the image structure. The system knows that hair is located at the top and in the back of the image, the neck is located in the lower part, and the part of the face adjacent to the ear is located in the front part. The approximate size of the auricle falling into the image field is also known. Images themselves were formed with a size of 500 × 500 pixels, and the enclosing rectangular window was considered with a size of 200 × 360 pixels.

When the generated image was fed to the input of the system, pixel codes were determined in the selected image areas.

The entire array of numbers was divided into intersecting areas of 200 × 360 pixels. Each area is shifted one pixel vertically and horizontally. The hair pixel code and the body pixel code in the image were determined. In accordance with the partition of the image into rectangular areas 200 × 360, an array of average values was formed for each rectangular area. Average values are in the upper left pixel of the numerical array of each area.

In accordance with the obtained array of average values, an array of values was formed according to the following formula:

$$W_i = \begin{cases} 1, \text{ if } \left|V_{i,\text{mean}} - A_{\text{mean}}\right| \leq F \\ 0, \text{ in other case} \end{cases}, \tag{3}$$

where $V_{i,\text{mean}}$—average value of the corresponding i-th window;

$A_{\text{mean}}$—the selected value of the average value, which corresponds to the pixel codes of the array of average values and located on the edges of the image of the auricle;

F—selected confidence interval value.

To select $A_{\text{mean}}$ and F additional algorithms are used that are aimed at their exact definition.

The $A_{\text{mean}}$ value is calculated by determining the value of the code of the human hair in the image and the value of the code defining the characteristics belonging to the auricle. The average value of these two values of the image arrays allows you to determine as accurately as possible $A_{\text{mean}}$. The value of F is calculated in accordance with statistical data obtained experimentally.

On each obtained image, the codes of the auricles are different (determined), as well as the codes of the hair are also different. Therefore, the system itself must determine them, taking into account their location in the image. Figure 8 presents examples of the selected parts of the image according to the described algorithm.

As a result of the algorithm, a different number of images is allocated in accordance with a certain $A_{\text{mean}}$. After binarization, images are obtained that display the binary ear in whole or in part. All of these images are processable. First, black pixels are removed, and small groups of black pixels are also deleted. To do this, the cellular automata (CA) are used that implement the local logical function AND and one of the classical forms of neighborhood (von Neumann neighborhood or Moore neighborhood) [14, 15]. As a result of the first step, the edge black pixels of each group switch to the logical "0" state. After several such operations, small groups of

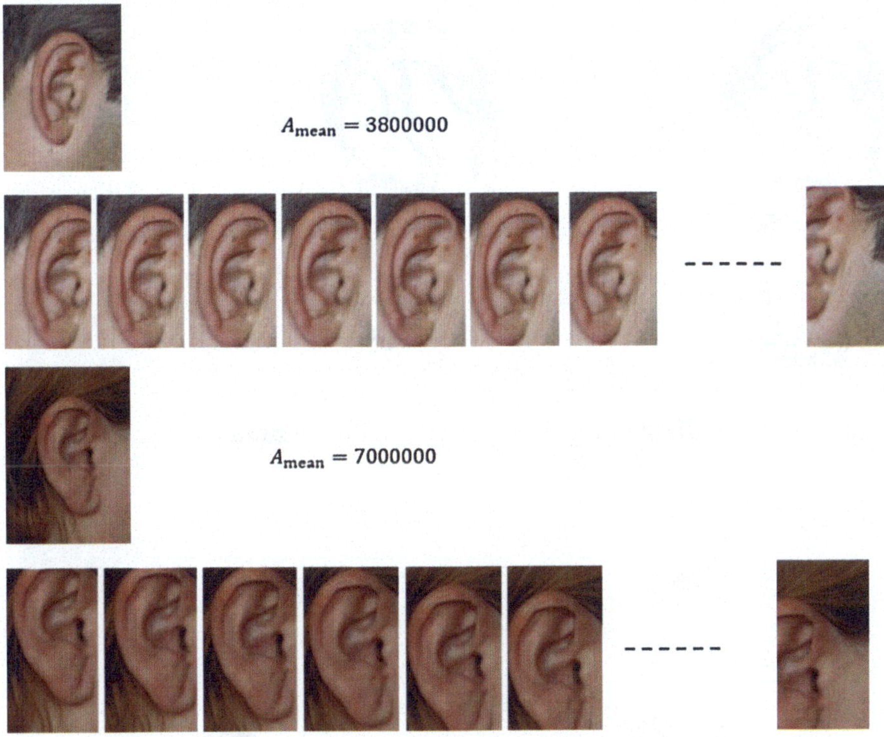


**Fig. 8** Examples of images of selected windows when $A_{mean} = 3800000$ *and* $A_{mean} = 7000000$

**Fig. 9** An example of an image after applying the theory of CA

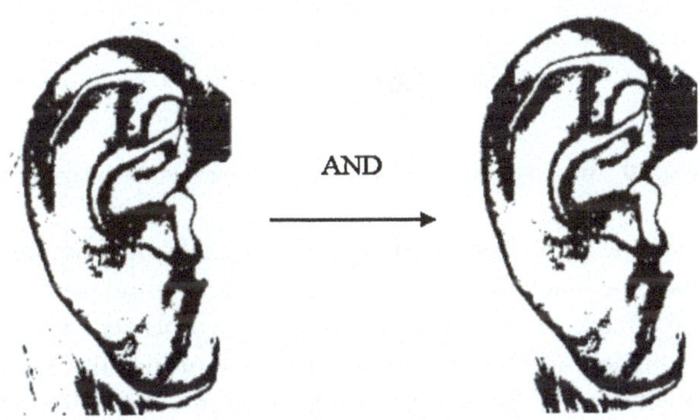


black pixels are deleted (Fig. 9). To implement such an operation, an elementary CA is often used for each row and column of the image matrix, depending on the location.

Removal of other parts not belonging to the auricle image is carried out using the Radon transform [11–13]. The projections of the Radon transform are formed and the projections formed in the directions of $30^0$, $60^0$, $120^0$, and $150^0$ are analyzed and the excess parts of the image are determined. If there are edge parts of the projection

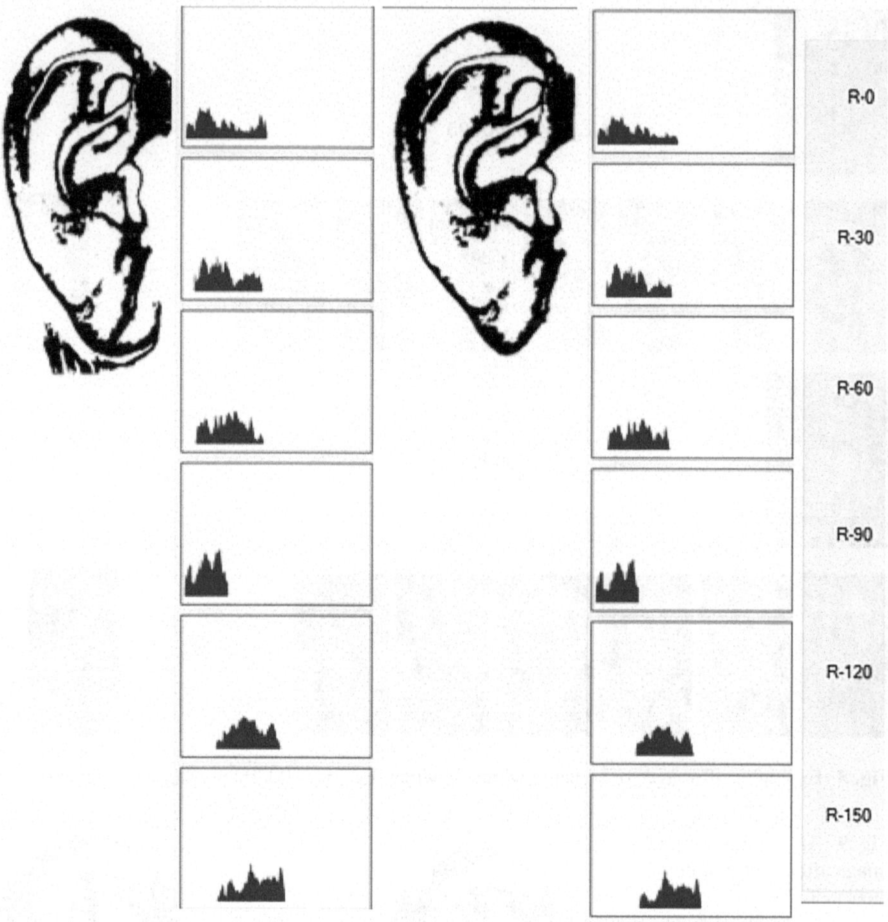


**Fig. 10** An example of removing excess pixels in the image of the auricle

of the Radon transform that are separated from the main whole projection of the ear, then the black pixels in this area become white pixels. Also, these projections identify black pixels located in the lower corners of a rectangular window, and are redundant (do not belong to the image of the auricle). Figure 10 shows an example of images of the auricle and projections of the Radon transform before and after removing extra pixels.

Also, according to the projections of the Radon transform, the location of the ear image in the rectangular window was determined and only those images of the auricle were selected, in which the projections of the Radon transform were located at a certain distance from the edge. As a result of such actions, the number of images of one auricle decreased.

The resulting images formed the standards, and also fed to the input of the identification system. The previously described method is not acceptable for such images, but the identification method based on the Radon transform is well used. Identification is carried out by forming six projections of the Radon transform in the directions $0^0$, $30^0$, $60^0$, $90^0$, $120^0$, and $150^0$. For each image, such projections of the Radon transform are different from each other; however, they coincide with the reference projections on the selected confidence interval with a certain threshold percentage of coincidences. Based on the results of the comparison, a decision is made on identification.

This method allows you to identify a person at a certain distance and can be device independent.

## 5 Conclusion

The proposed method and the developed software allows to identify a person with high accuracy due to the proposed division of the geometric shape of the auricle into six areas according to certain criteria. Most of these characteristic features differ for each ear, as well as their relationship. The method is simple to implement; however, it requires a special arrangement of the head of the identified person, as well as special tools to remove the remaining parts of the human body in the image. The method gives good values of FRR and FAR at confidence intervals of 0.2, 0.3, and 0.8 with the coincidence of 4, 5, and 6 parameters.

In addition, an effective method for extracting the auricle in a complex image, in which other parts of the body adjacent to the ear are present, is proposed. Effective selection of averages for all image areas, the use of the theory of cellular automata and the theory of Radon transform allowed us to isolate the image of the auricle and not use those initial images in which the image of the ear was not fully displayed.

In future works, it is planned to improve the algorithm for automatically extracting the ear image on a complex image obtained at a distance from a person. It is also planned to conduct research to search for characteristic features in the image of the ear, which would give a high degree of distinction and individuality.

## References

1. Burge, M., & Burger, W. (1998). Ear biometrics. In A. Jain, R. Bolle, & S. Pankanti (Eds.), *Biometrics: Personal identification in a networked society* (pp. 273–286). New York: Kluwer Academic.
2. Hoogstrate, A.J., Van den Heuvel, H., Huyben, E. (2000). *Ear identification based on surveillance camera's images*. Version updated May 31, 2000. Retrieved October 7, 2003, from: http://www.forensic-evidence.com/site/ID/IDearCamera.html
3. Kumar, A. (2008). Incorporating cohort information for reliable palmprint authentication. In *Sixth Indian Conference on Computer Vision, Graphics & Image Processing* (pp. 583–590). Los Alamitos, CA: IEEE Computer Society.

4. Nejati, H., Zhang, L., Sim, T., Martinez-Marroquin, E., & Dong, G. (2012). Wonder ears: Identification of identical twins from ear images. In *Proceedings of the international conference on pattern recognition* (pp. 1201–1204). New York: IEEE.
5. Xu, X., Mu, Z., & Yuan, L. (2007). Feature-level fusion method based on KFDA for multimodal recognition fusing ear and profile face. In *Proceedings of the international conference on wavelet analysis and pattern recognition* (pp. 1306–1310). Piscataway, NJ: IEEE.
6. Afolabi, A., & Ademiluyi, D. (2015). Ear symmetry determination in ear based human recognition. *Institute of Ear Biometric Research, 5*(6). http://www.iebresearch.org/1194765-88.pdf.
7. Choraś, M. (2005). Ear biometrics based on geometrical feature extraction. *Electronic Letters on Computer Vision and Image Analysis, 5*(3), 84–95.
8. Wahab, N. K. A., Hemayed, E. E., & Fayek, M. B. (2012). HEARD: An automatic human EAR detection technique. In *International conference on engineering and technology (ICET)*. Piscataway, NJ: IEEE.
9. Darmstadt, H. (2014). Effects of severe signal degradation on ear detection. In *2nd international workshop on biometrics and forensics* (Vol. 1, pp. 0–5). Piscataway, NJ: IEEE.
10. Jiang, J., Zhang, H., Zhang, Q., Lu, J., Ma, Z., & Xu, K. (2014). Ear feature region detection based on a combined image segmentation algorithm-KRM. *Proceedings of SPIE - The International Society for Optical Engineering, 8942*, 89420Z–89420Z–8.
11. Bilan, S., Motornyuk, R., & Bilan, S. (2014). Method of hardware selection of characteristic features based on radon transformation and not sensitive to rotation, shifting and scale of the input images. *Advances in Image and Video Processing, 2*(4), 12–23.
12. Belan, S. N., & Motornyuk, R. L. (2013). Extraction of characteristic features of images with the help of the radon transform and its hardware implementation in terms of cellular automata. *Cybernetics and Systems Analysis., 49*(1), 7–14.
13. Motornyuk, R.L. (2013). *Computer-aided methods for identifying images of moving objects based on cellular automata with a hexagonal coating*. Dissertation for the degree of candidate of technical sciences (UDC 004.932: 519.713 (043.3)), Kiev: SUIT, 2013.
14. Wolfram, S. (2002). *A new kind of science*/Wolfram Media.
15. Bilan, S. (2017). *Formation methods, models, and hardware implementation of pseudorandom number generators: Emerging research and opportunities*. Hershey, PA: IGI Global.

# Index





S. Bilan et al. (eds.), *Biometric Identification Technologies Based on Modern Data Mining Methods*, https://doi.org/10.1007/978-3-030-48378-4

The manufacturer's authorised representative in the EU is Springer Nature Customer Service Centre GmbH, Europaplatz 3, 69115 Heidelberg, Germany. If you have any concerns regarding our products, please contact ProductSafety@springernature.com

Printed and bound by CPI Group (UK) Ltd, Croydon, CR0 4YY

22/09/2026

02221060-0002